Terrain analysis and remote sensing

List of contributing authors

Alison Cook, Department of Geography, University of Sheffield

Peter J. Hancock, Department of Geography, University of Zimbabwe

John R. Hardy, Department of Geography, University of Reading

Christopher O. Justice, NASA/Goddard Space Flight Center, Greenbelt, Maryland, USA

Colin W. Mitchell, Department of Geography, University of Reading

John R. G. Townshend, Department of Geography, University of Reading

David F. Williams, Clyde Surveys, Maidenhead, Berkshire, UK

Terrain analysis and remote sensing

Edited by
John R. G. Townshend
Department of Geography, University of Reading

London
GEORGE ALLEN & UNWIN
Boston   Sydney

GEORGE ALLEN & UNWIN LTD
40 Museum Street, London WC1A 1LU

British Library Cataloguing in Publication Data

Terrain analysis and remote sensing.
  1. Natural resources – Remote sensing
I. Townshend, J R G
  333   HC55   80-41138

ISBN 0-04-551036-9
ISBN 0-04-551037-7 Pbk

Set in 10 on 12 point Times by Servis Filmsetting Ltd, Manchester
and printed in Great Britain
by Mackays of Chatham Ltd

To Professor Ronnie Savigear

As a token of appreciation for his efforts directed towards the scientific study of terrain and remote sensing

Preface

Remote sensing has been used in terrain analysis for several decades in the form of conventional black and white aerial photographs. During the past decade or so, newer forms of remote sensing and improved methods of data analysis have become widely available. A number of texts now exist outlining the principles of remote sensing and their applications (see Ch. 2) and a somewhat smaller number has been published reviewing the principles of terrain analysis (see Ch. 1). As its name implies the present book is focused on the frontier zone between these two subject areas.

Hopefully this book contains a more unified set of statements than the usual collection of essays bound in book-form: certainly this was our intention. Apart from our conscious efforts, two factors have helped achievement of this end. The authors have all spent a substantial part of their careers in a single university department, namely the Department of Geography in the University of Reading (with the exception of Alison Cook who spent only six months there) and, as this volume's dedication indicates, the authors owe much to the efforts of Professor R. A. G. Savigear who was primarily responsible for the development of remote sensing in the department.

The cohesiveness of the volume has also been aided by the editor having partially or wholly written several of the essays: the extent of this contribution largely arose from the commitments of others preventing completion of their intended chapters.

Nevertheless multi-authorship means that variations in approach are inevitable and these are fairly displayed in the case-studies which form the second half of this book. These are by no means completely representative of the possible applications predicated on the foundations provided by the first five systematic chapters: this could have been achieved only by a very much longer volume. Instead the case-studies are intended to indicate the range of possible applications primarily by drawing on the research by the individual authors.

Production of this book has relied on the efforts of many people apart from the authors. Photographic illustrations are largely the result of the skills of Philip Brice, formerly of Reading University and now at Manchester Polytechnic, and of Harry Walkland of Reading University. Line diagrams were prepared by Sheila Dance and Brian Rogers of the Geography Department, Reading University and, in the case of Chapter 9, by the drawing staff of the Department of Geography, University of Sheffield. Preparation of the typescript has been primarily the responsibility of Margaret Birch and Pam Dixon of the same department: they have coped with drafts, redrafts, rewritten versions of lost drafts and so much more. The patience and encouragement of the publishers and in particular of Roger Jones is gratefully acknowledged. Finally I wish to thank my wife, Jan, for her encouragement and forbearance during the editing of this volume.

John Townshend
Bowie, Maryland
4 November 1979

Acknowledgements

The authors and editor gratefully acknowledge the permission granted by the following to reproduce illustrative material and tables. Every effort has been made to contact organizations and individuals to obtain permission and we apologise if any have been omitted from this list. Numbers in parentheses refer to text figures unless otherwise specified:

Macmillan Publishing Co., New York (1.2); Commission of the European Communities, Joint Research Centre, Ispra, Italy (2.3, 2.4, 6.3, Plate 1, Tables 6.3 & 6.4); UK Ministry of Defence (2.7, 4.17, 5.2, 5.6, 10.10); Mr E. Milton (3.4b, c & d); Drs M. Barnett and P. Harnett, Optics Section, Department of Physics, Imperial College (4.12, 1.14b, 4.15); Dr R. M. Haralick (4.16); Institute of Electrical and Electronic Engineers, New York (4.22 and Table 4.3); Electromagnetic Systems Laboratory Inc., Sunnyvale, California (4.24); R.F. Buttery (5.1); Director General, Instituto Geografico Nacional, Spain (6.7); Director of Surveys and Mapping, Ministry of Lands, Housing and Urban Development, Tanzania (9.3, 9.4, 9.5); Dr J. Doornkamp (10.3); Food and Agriculture Organization of the United Nations (parts of Table 8.2); Unesco, for data used in Table 8.2 from *Arid lands: a geographical appraisal*, © Unesco 1966; Dr Y. J. Lee, Pacific Forest Research Center (4.1); Dr R. M. Hoffer, Laboratory for Applications of Remote Sensing (Table 4.4); Soil Conservation Service, USDA (Table 1.3); General Electric Inc. (Table 11.1); Springer-Verlag (parts of Table 1.2).

We are happy to acknowledge the source of the following non-copyright illustrations:

National Aeronautical and Space Administration (NASA) (2.5, 2.9, 3.3, 4.2, 4.3, 4.4, 4,7, 4.14a, 5.8, 6.1, 6.5, 6.8, 7.5, 8.2, 8.9, 9.6, 10.4, Plate 3, Tables 2.3 & 6.1); US Geological Survey, Department of the Interior (5.5).

Contents

xii

1 An introduction to the study of terrain

John R. G. Townshend

1.1 Introduction

Terrestrial remote sensing can broadly be defined as the set of techniques used to obtain information about the Earth's surface and atmosphere at some distance from them, usually by means of radiation from the electromagnetic spectrum.

One of the principal contributions of terrestrial remote sensing has been to increase our knowledge of the land surface and hence to improve its use. By way of an introduction to the rest of the book, in this chapter we review the need for information about land or terrain resources (terms which we use as synonyms), define some of the important concepts associated with terrain and, finally, outline the distinctive role that remote sensing plays.

In Chapter 2, the techniques of data acquisition by remote sensing are dealt with in some detail, followed by an account of how ground-level observations are integrated with remote sensing data (Ch. 3). Collecting the data is only the first stage, since they have to be analyzed to make them useful, and techniques for doing this are described in Chapters 4 and 5. Chapter 5 is particularly concerned with the creation of regions for the purposes of terrain or land evaluation. Various types of applications are presented in the second half of the book to demonstrate the value and limitations of remote sensing in contributing to the study of land.

Awareness of the varying qualities of terrain or land resources is as old as man himself; indeed the very definition of any terrain resource must always contain a significant cultural dimension. Conscious recognition of the benefits of sensible land management was explicitly described by Greek and Roman writers:

> The Greek and Roman treatises on agriculture from the *Oeconomicus* of Xenophon with his praises of Persian agriculture, to the *Natural history* of Pliny have the strong flavour of nature study, of watching and observing nature to learn the arts of sowing, tilling and plant breeding, while the writers of the Roman period like Varro, Columella and Pliny were deeply interested in the improvement of soils, methods of plowing, irrigation, drainage, removal of stones, clearing away of thickets, winning of new lands for cultivation, manuring and insect control. (Glacken 1967, p.13.)

Peoples without any written tradition have been observed in the 20th century to have a sensitive comprehension of the varying capabilities of terrain, even though their technology may be very simple (de Schlipre 1956, Watters 1960a, b) (Table 1.1). This is strongly indicative of comparable understanding by early man who was at similar technological levels, and an understanding of land or terrain capability undoubtedly formed one of the most important aspects of the acquired and transmitted knowledge of early man.

1

Table 1.1 Terrain terminology of the Azande (de Schlippe 1956).

Ri-ngbi	plateau
Kpakpangbere	bare, hardened ferricrete above 'break away'
Munga	bare, hardened ferricrete, not necessarily at 'break away'
Ri-di	gully with a permanent spring
Mbungunu	gully formed by natural erosion above a spring without permanent water
Genefukosa	gully started by a path on a slope
Barogodi	colluvial deposits in valley bottom caused by recent erosion
Mbudu-rago	locations with deep dark soils without stones
Ndawiri	marshy treeless river valley
Gbundu	tall rain forests
Ngaragba	grass woodland after bush fire
Ngasu	virgin land cultivated for first time
Bokuti	special type of grass fallow in which *Pennisetum pediculatum* (ngamu) is dominant grass

Even if a long-standing cognition of the significance of terrain has permeated man's activities for centuries, it was not until the 20th century that formal attempts were made to document land resources systematically, in ways which permit scientific predictions of their capabilities under various imposed duties. Knowledge of physical terrain resources forms only one of the many sets of factors which need to be considered in making decisions about the utilization and management of land. Economic, social, cultural and political factors can be of equal or, quite frequently, of greater significance. Nevertheless, increasing pressure on terrain resources as a result of both the rising world population and the higher per capita use of resources, particularly in the developed countries, means that plans must be devised to conserve and to improve renewable resources and to use non-renewable resources frugally. Such plans must be based on a knowledge of terrain resources, and this in turn demands the development of information systems for the efficient classification, compilation, mapping, storage, retrieval and usage of terrain resources information. The monitoring of changes in these characteristics is required especially for renewable resources, in order to assess the needs for changes in land use and management, and to monitor the impact of any changes which are instituted. Models designed to estimate the response of land to the imposition of new duties require the provision of data for their calibration and application to particular areas. The results produced at each of these stages will then be integrated with other factors to aid in the creation and implementation of planning procedures.

1.2 Crucial needs for terrain resources information

The various needs of man often make demands upon land, as is shown schematically in Figure 1.1. Provision of an adequate diet for the world's population at present and for the thousands of millions of additional people who will occupy this planet in the next few decades are the most important reasons for acquiring information about terrain resources. Confidence in the ability of the world to increase its food production varies substantially. Some people are relatively optimistic: 'We believe we can feed a world of 15 billion people 200 years from now with conventional food produced conventionally.' (Chou *et al.* 1977, p. 298.) Methods by which this can be done primarily involve the opening up of large areas of new land, increasing multiple cropping, raising yields by introduction of new high-yield varieties and increasing the use of fertilizers. Others are less optimistic: Crosson and Frederick (1977, p. 60) point in particular to the vulnerability of 'green revolution' techniques to external energy sources; Tinbergen (1977, p. 106) and Perelman (1977, p. 142) are doubtful about the extent and usefulness of existing unused areas and are pessimistic about the ability of many poor countries to increase yields because of a combination of physical limitations and adverse socio-economic factors. In part, these disagreements arise because of our lack of knowledge about existing terrain resources.

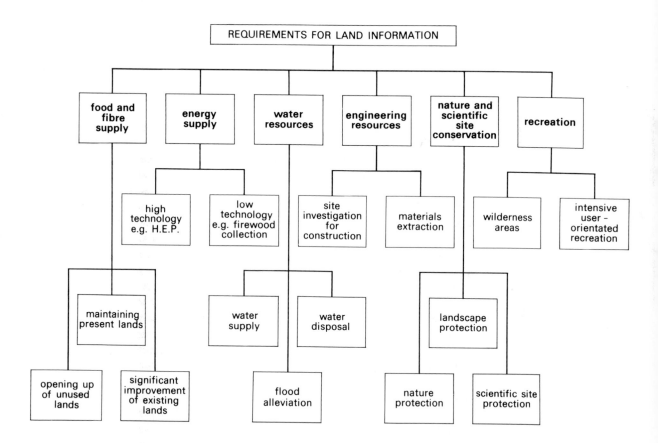

It is instructive to consider the changing appraisals of humid tropical lands. At first, they were seen as areas of enormous agricultural potential, primarily on the basis of their luxuriant vegetation, but this view was overtaken by much less optimistic appraisals, because of the many physical limitations. The latter apparently made it very difficult to match the natural biological productivity in agricultural terms (Gourou 1953, Goodland & Irwin 1975). But improvement in our knowledge of terrain resources and in agricultural practices have, over the past few years, greatly improved hopes of using much of these lands very productively (Buol & Sanchez 1978).

The relative importance of opening new lands and of intensification of existing land use for increasing food supplies has been changing progressively, as unused

Figure 1.1 Examples of needs for land resources information.

lands have declined in area. Many supposedly virgin lands, such as the Amazon rain forest and the Taiga of North America, are already used, albeit at a relatively low intensity. The Food and Agricultural Organization (FAO) expected the 1970s to be a period of absolute decline in the world's grazing lands. In part, this was because of overgrazing leading to the removal of land from the total stock. The mere maintenance of the productivity of existing lands, both for grazing and arable use, will remain a severe challenge in many regions, due to degradation by processes such as water and wind erosion, salinization and nutrient depletion (FAO 1971, 1977). Optimization of agricultural use of land is of continuing concern in developed countries and demands a knowledge of the biophysical capabilities of land (e.g. Mackney 1974, Bibby & Mackney 1969).

Many of the above points are relevant to agricultural production for non-food requirements (such as fibre for clothing) and are also apposite for forestry production: the time scales involved in the latter are, of course, much longer than those for agricultural production.

Wood is also a resource of a quite different kind, namely as a fuel, especially in developing countries. The shortage of firewood supplies is becoming an increasingly acute problem, especially in southern Asia (Frederick 1976), where it forms the principal domestic fuel and where forest reserves are being rapidly depleted. Moreover, denudation of woodland cover is leading to soil degradation and consequent downstream sedimentation with increases in flooding.

High-technology energy supply also makes demands on terrain information systems, primarily in assessing the environmental impacts associated with activities such as fossil-fuel extraction, the siting of power stations including hydro-electric ones, and the routing of transmission lines.

Requirements for terrain information in relation to the supply of water for domestic, agricultural and industrial uses include the impact of land cover changes on water supplies (Pereira 1973); the impact of new rural water supplies on land degradation (Rapp *et al.* 1972); the estimation of hydrological properties where direct observations are absent (Rodda 1969, Natural Environment Research Council 1975); and as a preliminary to hydrological investigations for locating surface and subsurface water supplies and estimating the need for irrigation water (Carson *et al.* 1977). The efficiency of waste disposal, especially from localized sources (as in cess pits) is also a function of terrain characteristics, and land-capability methods are regularly used to advise people on the location and design of such installations (Miller 1978, Galloway *et al.* 1975).

A knowledge of the physical properties of surface materials is important for many engineering activities (FAO 1973), such as the prevention of soil erosion and, hence, the maintenance of soil productivity, but it is also relevant in the context of road constructions (Thornburn 1966), which can easily be destroyed following the action of gullying and mass movements. The subject of slope stability has properly been the concern of civil engineers, who have developed sophisticated tests to assess slopes and their constituent materials, but the need for a more comprehensive evaluation of the local environments of roads has become increasingly apparent in preliminary reconnaissance studies of road alignment. Similarly, the location of suitable sites for buildings and other constructions may also benefit from a knowledge of the general land properties, as well as from detailed civil engineering techniques (Coates 1976). For example, the identification of sites in arid areas, where gypsum weathering of foundations is likely to occur, is of great significance (Brunsden *et al.* 1979). Other engineering needs include obtaining satisfactory construction materials from near-surface situations.

4

Engineering resources are related to the activities of military organizations (Parry *et al.* 1968, Grabau 1968), whose research institutions have been instrumental in the development of more quantitative approaches to terrain evaluation. Trafficability (or the ease with which terrain can be crossed) has formed the focus of much of the work, but the procedures involved in the construction of structures such as airfields have also been modelled (Benn & Grabau 1968).

The uses of land described above can all have an economic value assigned to them, though there are other sorts of values which may be relevant. For example, urban land use is often demonstrably more valuable in monetary terms than is agricultural land use, even on highly productive agricultural lands, but in Britain, the latter are specifically designated as such, to minimize their destruction.

Some uses of land have an apparently relatively minor economic justification; these include the protection of specific animals and plants, of sites of high scientific importance and of landscapes as a whole, whether natural or, more frequently, modified by man (Simmons 1974). The latter, in particular, can benefit from a systematic terrain analysis, in order that the full variety of landscapes may be conserved. Recreational uses can, at one extreme, have a very strong economic component or, at the other extreme, have a justification almost entirely non-economic, as in the case of designated wilderness areas, where development is expressly forbidden and a low density of visitors is deliberately maintained (Smith & Krutilla 1976). Most recreational uses fall somewhere between these two extremes. Assessing the ability of land to sustain the pressure of tourists, often in conjunction with complementary uses, clearly demands land information of various types (Montgomery & Edminster 1966).

Urban development also makes demands on terrain data. McHarg (1969) has stressed the importance of combining physiographic information with social values associated with wildlife, recreation and residence; a reduction in the impact of urbanization or suburbanization has also been stressed, especially in the United States (Beckman & Berg 1968). The most pressing problems of land utilization and management in the less developed countries are associated with increasing food, water and energy supplies. In contrast, many people in the developed countries are more concerned with environmental degradation, and the need to protect or conserve existing resources from the harm of changing impacts, though this ecocentric mode of thought may well be at odds with the inhabitants' technocentric existence (O'Riordan 1976).

No final pattern of land utilization is likely to be achieved in any area. 'In major degree, the practice of land use is a continuing search for some acceptable balance to this unchanging but inescapable people–land relationship. By their nature [such] balances are always unstable.' (Davis 1976, p. 285.) For the same reason, no complete, final collection of land resources data can ever take place. Furthermore, increases in our comprehension of the workings of the biophysical environment will themselves lead to requirements for new data.

1.3 Basic concepts of terrain

1.3.1 *Terrain and land*

Both **'terrain'** and **'land'** are words with everyday meanings closely related to their more precise technical definitions. Christian and Stewart (1968) provided a definition of land or terrain which stresses not only the range of factors to be included, such as climate, topography, soil, vegetation, fauna and water, but also the interactions between them which should be considered in characterizing land. On the basis of their work and refinements especially by Brinkman and Smyth (1973), the FAO has defined land as:

> . . . an area of the Earth's surface, the characteristics of which embrace all reasonably stable, or predictably cyclic attributes of the biosphere vertically above and below this area including those of the atmosphere, the soil and underlying geology, the hydrology, the plant and animal populations and the results of past and present human activity, to the extent that these attributes exert a significant influence on present and future uses of the land by man. (FAO 1976, p. 67.)

Whether intentional or not, this definition would appear to be too broad, both in terms of above- and below-surface properties. Relevant atmospheric characteristics must include the soil atmosphere and some consideration of above- but near-surface, atmospheric properties, in order that the Earth–atmosphere interface be included. However, inclusion of all those atmospheric properties which ultimately have a significant influence on land use would extend the zone under consideration at least up to the tropopause. Similarly, it is unreasonable to include more than a small proportion of subsurface geological characteristics. Additionally, given man's inadequate understanding of his physical environment, there seems little justification for including the qualification that the properties should be 'stable' or even 'predictably cyclic'. For these reasons, we prefer the following definition:

> *Land is an area of the Earth's surface which is characterised by a distinctive assemblage of attributes and interlinking processes in space and time of soils and other surface materials, their atmosphere and water, the landforms, vegetation and animal populations, as well as the results of human activity; also included to the extent that they directly influence the characteristics of the land under consideration are suprasurface properties of the atmosphere, subsurface geological characteristics and the nature of the immediately surrounding land and water.*

This definition is more restricted and corresponds more closely to the phenomenon that is studied by those operationally concerned with terrain evaluation. Furthermore, it stresses the dynamic character of land, which is especially appropriate since one of the principal reasons for all land investigations is the spatial and temporal variability in the response of terrain properties to changes in external influences. We have not restricted the definition to include only attributes relevant to present and future uses of the land since this is more appropriately done in defining land qualities and capability, in Section 1.3.4. Finally we have included the characteristics of surrounding land to the extent they have a direct influence on the land under consideration.

We follow Christian and Stewart (1968) in treating 'land' and 'terrain' as synonyms, and the various authors of this book use either or both words according to personal preference.

1.3.2 *Terrain resources*

Definition of the term 'natural resource' has changed with improved understanding of man's relationships with his physical environment (Becht & Belzung 1975).

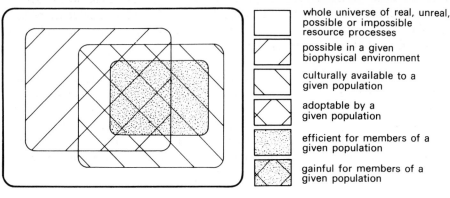

Sets of resource processes

whole universe of real, unreal, possible or impossible resource processes

possible in a given biophysical environment

culturally available to a given population

adoptable by a given population

efficient for members of a given population

gainful for members of a given population

Figure 1.2 Set depiction of a combined ecological, economic and ethnological approach to resources (from Firey 1960).

Originally, the term as applied to physical resources was used simply to describe certain tangible components of the physical environment. Zimmerman (1951, p. 7) argued for a replacement of this concept with an 'operational or functional theory of resources':

> The word 'resource' does not refer to a thing or a substance but to the function which a thing or a substance may perform or to an operation in which it may take part, namely the function or operation of attaining a given end such as satisfying a want. In other words the word 'resource' is an abstraction reflecting human appraisal and relating to a function or operation.

He is somewhat inconsistent, since in the remainder of his mammoth text resources are treated as tangible objects, and chapter titles such as 'fibers', 'trees and tree crops' are included.

Firey (1960) argued that there are three principal approaches to resource phenomena: an economic one based on scarcity and gain; an ecological one based on equilibrium, stability or balance; and an ethnological one based on what is adoptable by a given group. He thus identified resources or 'resource complexes' in terms of an overlapping series of sets (Fig. 1.2).

In the broadest sense, **terrain resources** include all those features and processes of land which can be used in some way to fulfil certain human needs (Vink 1975). Using Firey's concepts of natural resources, we can define land resources more formally as follows:

> The terrain resources of a locality are those subsets of its biophysical attributes whose use is beneficial to man: such usefulness can be assessed both according to economic and non-economic criteria. Even if the former is pre-eminent, due consideration must be paid to certain social, political and ethical standards; if non-economic criteria prevail, then use of the attributes must nevertheless in some sense be affordable. Moreover, the usefulness of land resources is predicated on the availability of a requisite level and type of technology.

This definition has several important implications for our understanding of the nature of terrain resources. First, we note that they are not fixed immutable parts of the environment. Changes in economic, social, political, ethical or technological factors will lead both to the creation of new resources and to the disappearance of others. A corollary of this is that existing terrain resources may change in value 7 through time. Secondly, it follows that the same piece of land may be possessed simultaneously of several resources: indeed this is the norm rather than the exception. The actual use or uses of the land will normally arise out of the resolution of several complex conflicting claims. A final implication is that although we might just conceivably produce a definitive lasting statement of a region's terrain characteristics, no such final document can even conceptually be formulated for terrain resources.

A distinction should be made between **renewable** and **non-renewable resources**. Terrain resources assessment and management are frequently concerned with the maintenance of the productivity or usefulness of land, but the extraction of surface materials, usually for engineering construction purposes, obviously involves inherently non-renewable resources.

1.3.3 *Land cover and land use*

These two terms are often used almost interchangeably, but this is really quite improper. **Land cover** describes 'the vegetational and artificial coverings of the land surface' (Burley 1961), and thus forms an attribute of land or terrain. In contrast, land use includes 'man's activities on land which are directly related to the land' (Clawson & Stewart 1965). Given that we are particularly concerned with observable properties of the land surface, it is worth noting how many other elements are encompassed in the definition of rural land use by reference to Table 1.2. Obviously, many of these elements would not normally be included in definitions of land or terrain.

Table 1.2 Attributes of rural land use types (Vink 1975, FAO 1976).

1. *Produce*, including goods (e.g. crops, livestock, timber) services (e.g. recreational facilities) or other benefits (e.g. wild life conservation)
2. *Market orientation*, including whether towards subsistence or commercial production
3. *Capital intensity*, including both initial long-term inputs and short-term animal production inputs
4. *Labour intensity*
5. *Power sources* (e.g. man's labour, draught animals, machinery using internal or external power sources)
6. *Technical knowledge*, including management skills
7. *Technology employed* (e.g. implements and machinery, fertilizers, livestock breeding, farm transport, methods of timber felling)
8. *Attitudes of users* (particularly value-system adopted)
9. *Infrastructure requirements* (e.g. saw mills, tea factories, advisory services)
10. *Size and configuration of land holdings*, including whether consolidated or fragmented
11. *Land tenure*, the legal or customary manner in which rights to land are held, by individuals or groups
12. *Income levels*, whether expressed *per capita*, or per unit of production (e.g farm) or per unit area

1.3.4 *Terrain characteristics, terrain qualities, and land capability*

Following the distinction made by Kellogg (1961) between soil characteristics and soil qualities, a similar distinction can be drawn between **terrain characteristics** and **terrain qualities**, the former referring to directly observable and measurable

8

properties, and the latter to terrain conditions which determine the degree to which a certain tract of land can be put to a certain use (Vink 1975). Terrain qualities such as productivity, fertility and erodibility are not observable or directly measurable but are complexes derived or estimated from terrain characteristics. Beneficial aspects of terrain qualities may sometimes therefore be equated with terrain resources.

Land capabilities or **suitabilities** describe the fitness of terrain to perform a defined use (FAO 1976, p. 17). Views concerning the precision with which these uses should be specified vary considerably. The land capability classification of the Soil Conservation Service of the US Department of Agriculture is based on a grouping of mapped soil units: 'primarily on the basis of their capability to produce common cultivated crops and pasture plants without deterioration over a long period of time'. (Klingebiel & Montgomery 1961.) Broadly the eight types of classes vary from land with soils having virtually no restrictions concerning their use, through classes with increasingly severe limitations, with a correspondingly narrower range of land-use possibilities (Table 1.3). Similar schemes have been adopted in several other countries (Olson 1974). Although generalized, the scheme is based on suitability for agriculture mainly, and forestry or other land uses would require a different scheme. Others would argue that suitability or capability should be quite specific: '. . . land should be rated on its value (suitability) for a specific purpose, since there is no absolute and generally applicable value of land'. (Beek & Bennema 1974, p. 60.) An even stronger view is that proposed by Benn and Grabau (1968) who, in the context of military land uses, state that the establishment of a mathematical formulation relating an activity's performance to terrain properties is an absolutely necessary precursor to terrain classification.

It has been argued that the parametric approach as described by Benn and Grabau (1968) is impractical for most purposes, because the demands made on data are too great (Aitchison & Grant 1968). Less stringent parametric methods, often based on more empirically derived relationships, have been used especially for agricultural purposes (Olson 1974, Riquier 1974), with strong reliance being placed on climatic data (Garbouchev et al. 1974, Teaci & Burt 1974).

Although general purpose determinations of land capability may continue to be of value in some areas, the need to relate land capability assessments more

Table 1.3 Land capability classes (Klingebiel & Montgomery 1966).

Class
- I Soils with very few limitations restricting their use
- II Soils with some limitations reducing the selection of crops, or requiring moderate conservation practices
- III Soils with severe limitations reducing the choice of plants and/or requiring conservation practices
- IV Soils with very severe limitations restricting choice of plants and/or requiring very careful management
- V Soils with little or no erosion hazard, but use restricted largely to pasture, range, woodland or wildlife food and cover due to limitations impractical to remove
- VI Soils with very severe limitations, restricting use to pasture or range, woodland or wildlife
- VII Soils with very severe limitations restricting use to grazing, woodland or wildlife
- VIII Soils and landforms with such severe limitations that all commercial plant production is precluded and use is restricted to recreation, wildlife or water supply

Basic concepts of terrain

9

specifically to certain uses is undoubtedly growing and has clear implications for data-gathering methods. Terrain qualities that need to be estimated will change with improvements in the models relating land capability to terrain characteristics or with different demands being made on land. New properties may need to be estimated or the precision of the measurements will need to be improved. It is unrealistic therefore to expect that the collection of terrain characteristics will be a once-and-for-all exercise, though the creation of a comprehensive geographical information system should lead to a decline in the amount of data gathering. Advances in data-collection methods will themselves have an impact on land-capability assessments, since they will permit improvements in the use of existing models and encourage the development of new ones.

Improvements in land productivity are often relatively slow with progressive advances in land management, though more rapid changes can arise, for example, because of the introduction of new plant varieties, or new technology. The investment of large amounts of capital can transform an agricultural system and result in enormous increases in productivity, as in irrigation schemes, drainage of wetlands, and reclamation of land from the sea. The magnitude of the improvements has to be set against the high investments. Clearly the results of terrain analyses and evaluations necessary for such high investment projects will be very different from those for relatively low-cost improvements, and hence Vink (1975) usefully distinguishes between **potential** and **actual land suitability** respectively.

1.3.5 *Terrain analysis and evaluation*

Terrain or **land analysis** is the set of activities which leads to the compilation of terrain characteristics or terrain qualities. Terrain or land evaluation uses the characteristics and/or qualities extracted by terrain analysis, along with other properties, to assign a value to a piece of land, expressed either by a numerical value or by a judgement of its worth in qualitative terms (Mitchell 1973). In recent years valuable work has been devoted to specifying the results of terrain evaluation in economic terms (Vink 1975, Dudal 1978). Such an effort will clearly often be of great value especially in ensuring that land information is used more fully in regional and national planning. Nix (1968) even writes that: 'land evaluation simply means assigning a value to a specified unit area of land. In practice, the final expression of this value will be in economic terms'. (Nix 1968, p. 77.) This provides an unnecessarily restrictive definition of evaluation, since although an economically assigned value will almost always be useful, there are several other criteria which could be used. Recreational use of land may well demand an evaluation of its aesthetic value; social or political factors may alter priorities in land use despite the economic assessments that were made.

Terrain evaluation surveys therefore involve a variety of approaches from highly specific land suitability assessments to generalized assessments of the physical environment, such as the land systems surveys of the Australian CSIRO (Commonwealth Scientific Research and Industrial Research Organisation).

Excellent reviews of terrain evaluation procedures may be found in Mitchell (1973) and Young (1973).

1.4 The integrated approach

The definitions of terrain which are introduced in the previous section imply that terrain can itself be studied as an operational subject in preference to separate analytical investigations of its component parts. Justification for an integrated approach to terrain analysis and evaluation is threefold. First, there is accumulated scientific knowledge which has demonstrated that functional inter-relationships between terrain characteristics are often close. Such knowledge stems from many of the sciences which study the separate components of terrain, such as hydrology, pedology and geomorphology, but also from more holistic studies in ecology and in systems theory (Bennett & Chorley 1978). Secondly, the particular assemblage of characteristics and processes at a locality is often found to extend spatially for some distance, hence providing the possibility for regionalization. The resultant regions are sometimes based on functional inter-relationships between terrain charac- teristics and qualities but may also simply rely on empirically observed correspon- dences. If valid, these two justifications provide a third and wholly pragmatic one, namely that we can hope to use a limited number of properties and observations to estimate many other characteristics and qualities. This is likely to be more cost- effective if only because fewer measurements need to be made, but it is more important for another reason. In any operational survey it is quite impossible to record land characteristics directly for anything but a small proportion of the available land. This is true of reconnaissance surveys (at scales of $> 1:250\,000$), but is also valid for very large scale surveys (at scales of $1:1000$); for example, the intensive testing of soils is inevitably carried out on very small samples in relation to the population of materials they represent. Therefore, we need the capability to make extrapolation throughout surrounding areas from highly local observations. To do this with no other data is possible by direct mathematical or statistical methods of surface fitting. But if we have spatially comprehensive information about a limited number of characteristics for the whole survey area we can hope to estimate other characteristics to a much higher degree of refinement. This is essentially why remotely sensed data, and to a lesser extent topographic maps, are so important in any terrain analysis or terrain evaluation.

Although this pragmatic justification of an integrated approach and the theoretical models of relationships between terrain characteristics are closely related, it is important to recognize their conceptual independence, otherwise there is grave danger of arguing circularly for the sake of expediency. The nature of the interdependency between terrain characteristics needs to be reviewed continuously on the basis of advances in systematic scientific studies and on the basis of field tests in a variety of environments. The complexity of inter-relationships of terrain properties means that establishment of a consistent relationship in one area cannot necessarily be extended to another.

1.5 The distinctive contribution of remote sensing in providing terrain resources data

So far we have established the needs for terrain resources information, introduced the basic concepts of terrain and the need for an integrated approach in terrain evaluation. During this discussion, the role of remote sensing has been alluded to several times. Details of remote sensing data collection and analysis are provided in Chapters 2 to 5, but it is useful now to draw together the basic factors underlying the distinctive contribution of remote sensing, largely using the terminology of Steiner (1972).

First, and most importantly, is the **spatial comprehensiveness** of remote sensing data compared with data collected at ground level. Thus even properties which can

11

be recorded more precisely at ground level are instead often inferred from remotely sensed data. As a result of this spatial comprehensiveness, **accessibility** is greatly improved, especially where survey areas are large or difficult to traverse. **Flexibility of aggregation** is also made relatively simple by the same characteristic, thus enabling almost any size or shape of area to be selected, including any administrative unit. Areal coverage can be obtained rapidly by remote sensing methods, in particular from space altitudes, ensuring **comparability of data** from different areas. Images from space altitudes are also of increasing significance in providing more **synoptic views** of large areas, which make broad-scale spatial variations of land properties more readily detectable. Data collection at ground level is certainly not eliminated by the use of remote sensing, but it can greatly improve the efficiency with which ground data collection can be performed (Ch. 3). For example, time spent on field work and travelling between sites can be reduced, thereby lowering overall personnel requirements. Repetitive imaging makes **monitoring** feasible, whether to highlight forthcoming problems, or to assess the success of measures taken to maintain or improve the use of land. This is greatly aided by the **historical record capability** of remote sensing images, namely that new data can often be extracted from older images collected for quite different purposes.

Extraction of land characteristics from remote sensing data may be relatively straightforward if they are land-cover or morphological features, or surface-material characteristics where land cover is sparse, since they are directly manifested. Most other characteristics, such as geological, hydrological, geomorphological and pedological ones, frequently have little or no direct effect on the imagery, but their indirect effect through the surface characteristics may well allow them to be inferred. Thus the value of remote sensing data is strongly dependent on the ability of the interpreter or machine-assisted systems (Ch. 4). The usefulness of inferences based on the surface appearance of imagery is an important justification for an integrated approach in land resources survey (Nunnally 1969).

Although black-and-white air photography will remain highly important, there are several significant properties of remote sensing techniques which are offering new opportunities for extracting terrain data. These include use of other parts of the electromagnetic spectrum which increases the range of terrain characteristics about which we can hope to extract information. Imagery from different altitudes (especially higher space platforms) is increasing spatial comprehensiveness and improving synoptic coverage. Satellite images also have the potential to improve the temporal frequency of observation, which enhances monitoring capabilities.

Finally, the limitations of remote sensing data are not ignored in this book. These include the lack of penetrating ability of most sensors, operational restrictions for most sensors due to weather conditions, problems of extracting the relatively small quantity of useful data from the enormous volume of data in an image and the need for personnel trained in new techniques. Although the authors are all confident of the role of remote sensing in land resources survey, an uncritical approach is not and should not be adopted. Remote sensing adds to our abilities; it does not replace traditional ones, though it may often enhance them.

We look to remote sensing to aid us in the compilation of base-line information, to monitor changes in terrain characteristics and ultimately to provide inputs for the modelling of terrain resources so that rational forecasts can be made of their future status. In this way plans can be formulated for the sensible usage, management and conservation of land resources.

12

References

Aitchison, G. D. and K. Grant 1968. Terrain evaluation for engineering purposes. In *Land evaluation*, G. A. Stewart (ed.), 125–46. Melbourne: Macmillan.

Becht, J. E. and L. D. Belzung 1975. *World resource management*. Englewood Cliffs, NJ: Prentice-Hall.

Beckman, N. and N. A. Berg 1968. *Soil, water and suburbia*. Proceedings of conference sponsored by USDA and USD of Housing and Urban Development, Jan. 1967, Washington, DC.

Benn, B. O. and W. E. Grabau 1968. Terrain evaluation as a function of user requirements. In *Land evaluation*, G. A. Stewart (ed.), 64–76. Melbourne: Macmillan.

Bennett, R. J. and R. J. Chorley 1978. *Environmental systems: philosophy, analysis and control*. London: Methuen.

Bibby, J. S. and D. Mackney 1969. *Land use capability classification*. Soil survey technical monograph, 1. Harpenden: Soil Survey of England and Wales.

Brinkman, R. and A. J. Smyth 1973. *Land evaluation for rural purposes*. Publ. 17, Int. Inst. Land Reclamation and Improvement, Wageningen.

Brunsden, D., J. C. Doornkamp and D. K. C. Jones 1979. The Bahrain surface materials resources survey and its application to regional planning. *Geogl J.* **145**, 1–35.

Buol, S. W. and P. A. Sanchez 1978. Rainy tropical climates: physical potential, present and improved farming systems. *Proc. 11th Conf. Int. Soc. Soil Sci., Edmonton, Canada* **2**, 292–310.

Burley, T. M. 1961. Land use or land utilization? *Prof. Geographer* **13**, 18–20.

Carson, E. E., R. Z. Wheaton and J. V. Mannering 1977. *Irrigation of field crops in Indiana*. West Lafayette, Indiana: Purdue University.

Chou, M., D. P. Harmon, H. Kann and S. H. Wittwer 1977. *World food prospects and agricultural production*. New York: Praeger.

Christian, C. S. and G. A. Stewart 1968. Methodology of integrated surveys. *Proc. Conf. Aerial Surveys and Integrated Studies, Toulouse, Unesco*, 233–80.

Clawson, M. and C. L. Stewart 1965. *Land use information. A critical survey of US statistics including possibilities for greater uniformity*. Baltimore, Md: Johns Hopkins University Press for Resources for the Future.

Coates, D. R. 1976. *Geomorphology and engineering*. Stroudsburg, Pa.: Dowden, Hutchinson & Ross.

Crossen, P. R. and D. Frederick 1977. *The World food situation*, Washington, DC: Resources for the Future.

David, K. P. 1976. *Land use*. New York: McGraw-Hill.

de Schlippe, P. 1956. *Shifting cultivation in Africa: the Zande system of agriculture*. London: Routledge & Kegan Paul.

Dudal, R. 1978. Land resources for agricultural development. *Proc. 11th Conf. Int. Soc. Soil Sci., Edmonton, Canada* **2**, 314–40.

FAO 1971. Land degradation. *Soils Bull.* no. 13.

FAO 1973. Soil survey interpretation for engineering purposes. *Soils Bull.* no. 19.

FAO 1974. Approaches to land classification. *Soils Bull.* no. 22.

FAO 1976. A framework for land evaluation. *Soils Bull.* no. 29.

FAO 1977. Assessing soil degradation. *Soils Bull.* no. 34.

Firey, W. 1960. *Man, mind and land*. Glencoe, Illinois: The Free Press.

Frederick, K. D. 1976. Energy use and agricultural production in developing areas. In *Changing resource problems of the fourth world*. R. J. Ridker (ed.), 90–123, Baltimore, Md: Johns Hopkins University Press.

Galloway, H. M., J. E. Yahner, G. Srinivasan and D. P. Franzmeier 1975. *User's guide to the general soils maps and interpretative data for counties of Indiana*. West Lafayette, Indiana: Purdue University.

Garbouchev, I., H. Trashliev, S. Krastonov, M. Elgabaly, M. L. Dewan and T. Okuno 1974. Land productivity in Bulgaria. In FAO (1974), 83–95.

Glacken, C. 1967. *Traces on the Rhodian shore*. Berkeley: University of California Press.

Goodland, R. J. A. and H. S. Irwin 1975. *Amazon jungle: green hell to red desert?* Amsterdam: Elsevier.

Gourou, P. 1953. *The tropical world*. London: Longman.

Grabau, W. E. 1968. An integrated system for exploiting quantitative terrain data for engineering purposes. In *Land evaluation*, G. A. Stewart (ed.), 211–20. Melbourne: Macmillan.

Kellogg, C. E. 1961. *Soil interpretation in the soil survey*. Washington DC: USDA Soil Conservation Services.

Klingebiel, A. A. and P. H. Montgomery 1961. *Land-capability classification*. Soil Conservation Service, Agriculture Handbook 210, Washington, DC.

Mackney, D. (ed.) 1974. *Soil type and land capability*. Soil survey technical monograph 4. Harpenden: Soil Survey of England and Wales.

McHarg, I. L. 1969. *Design with nature*, New York: Natural History Press.

Miller, F. P. 1978. Soil survey under pressure: the Maryland experience. *J. Soil Water Conserv*, **22**, 104–11.

Mitchell, C. W. 1973. *Terrain evaluation*. London: Longman.

Montgomery, P. H. and F. C. Edminster 1966. Use of soil surveys in planning for recreation. In *Soil surveys and land use planning*, L. J. Bartelli *et al.* (eds), 104–12. Madison, Wisconsin: Soil Science Society of America, and American Society of Agronomy.

Natural Environment Research Council 1975. *Flood studies report*. London.

Nix, H. A. 1968. The assessment of biological productivity. In *Land evaluation*, G. A. Stewart (ed.), 77–87. Melbourne: Macmillan.

Nunnally, N. R. 1969. Integrated landscape analysis with radar imagery. *Remote Sensing of the Environment* **1**, 1–6.

Olsen, G. W. 1974. Interpretative land classification in English-speaking countries. In FAO (1974), 1–25.

O'Riordan, T. 1976. *Environmentalism*. London: Pion.

Parry, J. T., J. A. Heigenbottom and W. R. Cowan 1968. Terrain analysis in mobility studies for military vehicles. In *Land evaluation*, G. A. Stewart (ed.), 160–70. Melbourne: Macmillan.

Pereira, H. C. 1973. *Land use and water resources*. Cambridge: Cambridge University Press.

Perelman, M. 1977. *Farming for profit in a hungry world*. Montclair, NJ: Allanheld, Osmun and Co.

Rapp, A., D. H. Murray-Rust, C. Christiansson and L. Berry 1972. Soil erosion and sedimentation in four catchments near Dodoma Tanzania. *Geog. Annlr* **54A**, 255–318.

Riquier, J. 1974. A summary of parametric methods of soil and land evaluation. In FAO (1974), 47–53.

Rodda, J. C. 1969. The flood hydrograph. In *Water, earth and man*, R. J. Chorley (ed.), 405–18. London: Methuen.

Simmons, I. 1974. *The ecology of natural resources*. London: Edward Arnold.

Smith, K. V. and J. V. Krutilla 1976. *Structure and properties of a wilderness travel simulator: an application to the Spanish Peaks area*. Baltimore, Md: Johns Hopkins University Press.

Steiner, D. 1972. Remote sensing input to information systems. In *Geographical data handling*, R. F. Tomlinson (ed.), 541–631. International Geographical Union.

Teaci, D. and M. Burt 1974. Land evaluation and classification in East-European countries. In FAO (1974), 35–46.

Thornburn, T. H. 1966. The use of agricultural soil surveys in the planning and construction of highways. In *Soil surveys and land use planning*, L. J. Bartelli *et al.* (eds), 87–103, Madison, Wisconsin: Soil Science Society of America, and American Society of Agronomy.

Tinbergen, J. 1977. RIO *Reshaping the international order*. Rep. to the Club of Rome, New American Library: New York.

Vink, A. P. A. 1975. *Land use in advancing agriculture*. New York: Springer-Verlag.

Watters, R. F. 1960a. Some forms of shifting cultivation in the S. W. Pacific. *J. Trop. Geog.* **14**, 35–50.

Watters, R. F. 1960b. The nature of shifting cultivation – a review of recent research. *Pacific Viewpoint* **1**, 59–99.

Young, A. 1973. Rural land evaluation. In *Evaluating the human environment*, J. A. Dawson and J. C. Doornkamp (eds), 5–33. London: Edward Arnold.

Zimmerman, E. W. 1951. *World resources and industries*, revised edn. New York: Harper & Row.

2 Data collection by remote sensing for land resources survey

John R. Hardy

2.1 Introduction: ground sampling, extrapolation and remote sensing

A situation which is common to nearly all resources surveys is that the size of the area to be surveyed, or the amount of detail, or both, prevent the survey team covering the whole area on the ground in a satisfactory manner. In other words, the survey team has a sampling problem: sample sites have to be examined on the ground, and extrapolation has to be made from these sites to the remainder of the area.

Nowadays, these extrapolations are usually based on remotely sensed data. Remote sensing involves the determination of properties of objects without being in physical contact with them. For the purposes of Earth resources surveying, the definition may be narrowed to the estimation of terrain characteristics and qualities by the use of data acquired from aircraft or spacecraft.

The use of remotely sensed data in resources surveys, in the form of vertical stereoscopic black and white aircraft photographs, has been a standard practice for several decades now. Perhaps the best known early work is that of the CSIRO in the land system approach (Ch. 5). Such aircraft photography is now supplemented frequently by other forms of remotely sensed data, acquired by other more recently developed sensors, which may be mounted in aircraft or spacecraft, particularly satellites.

These sensors detect various signals from the Earth's surface. One main group of signals, and the group with which this book is almost entirely concerned, is **electromagnetic radiation**, of various wavelengths and properties, reflected or emitted from the surface. Another group of signals embraces those which have been used for some decades in geophysical prospecting, mostly concerned with gravitational and magnetic forces. The methods have not commonly been used in integrated land resources survey and are not discussed in this book. More detailed information on the content of this chapter can be found in particular in Reeves (1975), Lintz and Simonett (1976), and Swain and Davis (1978).

2.2 Sensors, platforms and electromagnetic radiation

Remotely sensed data are acquired by **sensors** carried in aircraft or spacecraft. The craft carrying the sensor is referred to as the **platform**. These sensors, for the most part, sense and record **electromagnetic radiation** reflected or emitted by the Earth's surface, subject to modification by the intervening atmosphere. The energy-transfer process of electromagnetic radiation is thus the **medium** by which the properties of the Earth's surface are sensed remotely, and an understanding of this process is therefore of fundamental importance. Reviews of the subject to various levels of sophistication and detail can be found in a large number of books (e.g. Monteith 1973, Sellers 1965, Reeves 1975).

Every object with a temperature above absolute zero (0 K) emits its own characteristic spectrum of electromagnetic radiation. As the temperature of a body increases, so does the amount of radiation emitted, in proportion to the fourth power of the absolute temperature (Stefan–Boltzmann law). Thus the electromagnetic radiation from the Sun, the largest and hottest body near to the Earth, with its surface temperature of approximately 6000 K, dominates the energy relations of the Earth's surface. Another property of electromagnetic radiation is that the wavelength of peak emission decreases as the temperature of the emitting body increases (Wien's law). The peak radiation emission wavelength of the sun is at approximately 0·5 μm, in the green band of visible light, while the maximum emission wavelength of the Earth with a mean surface temperature of about 290 K is at approximately 10 μm, in the thermal band. Figure 2.1 illustrates the spectra of Sun and Earth, the former as it strikes the top of the atmosphere and the latter as it is emitted from the surface. Note that there is little overlap between the two curves.

Electromagnetic radiation can pass freely through the vacuum of space, but in contact with matter the energy may be transmitted, absorbed, scattered or reflected.

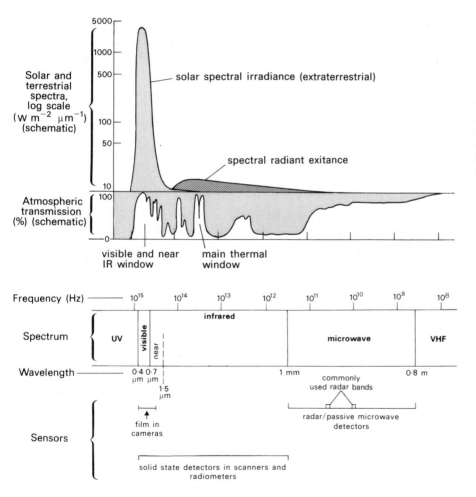

Figure 2.1 Schematic representation of salient features of the electromagnetic spectrum and sensors used in remote sensing. The boundaries between parts of the electromagnetic spectrum do not mark abrupt changes in the physical properties of the radiation. Several of the boundaries have no universally accepted definition.

1 μm (micrometre) = 1000 nm (nanometres) = 10^{-6} m (metres)

Absorbed energy may be re-emitted as radiation of a different wavelength governed by the properties of the absorbing and emitting body. Thus it is that solar energy absorbed at the Earth's surface is re-emitted from the Earth at much longer wavelengths as thermal energy. That part of integrated land resources survey concerned with remote sensing involves the determination of terrain properties from a study of electromagnetic radiation received and recorded at a sensor or sensors. Success depends almost entirely on a knowledge of the distribution of this radiation among the various components of transmission, absorption, scattering, reflection and emission, as it passes through the various elements of the Earth–atmosphere system.

The atmosphere is an important element in remote sensing studies, in two ways. The first is the 'nuisance value' of the atmosphere in a study of the Earth's surface, and the second is the large 'constructive' element of determination of atmospheric properties from remotely sensed data (Farrow 1975).

Cloud is the most obvious obstruction to Earth surface observation. Aerosol particles, dust and moisture are less obvious but nonetheless a persistent problem, scattering electromagnetic radiation. At some wavelengths, however, the atmosphere is opaque to electromagnetic radiation, for example in the approximate ranges 6–8 μm, and 14–17 μm, with the areas between, notably in this case 8–14 μm, being termed atmospheric 'windows' (Fig. 2.1).

Certain gases present in the atmosphere in small and variable amounts, such as ozone, carbon dioxide and water vapour, mainly affect the passage of radiation in these absorption bands of known wavelengths. Judicious use of sensors detecting radiation in these absorption bands can be made in atmospheric studies to determine the amounts of such gases present, and remote sensing techniques can also be used, for example, to increase atmospheric knowledge by the determination of temperature profiles. Such studies are relevant to land resources surveys only insofar as climatic factors affect land productivity, for example. Interested readers may refer to such works as Barrett (1974), while more general considerations of atmospheric physics are dealt with by Fleagle & Businger (1963).

Remote sensing systems in common use record radiation in various parts of the total electromagnetic spectrum, the visible, near-infrared, far-infrared or thermal, and the microwave, covering wavelengths from about 0·4 μm to 1 m (Fig. 2.1).

2.3 Types of sensors

Remote sensing systems are designed to record radiation in one or more parts of the electromagnetic spectrum. There are four main types of imaging sensors commonly used in remote sensing, which are described in more detail below. **Photographic** systems use cameras and film, and can record radiation only in the visible and parts of the near-infrared section of the spectrum. **Linescan** systems are used for a much wider range of wavelengths, the part of the spectrum sensed by a particular linescan system being dependent on the nature of the detector used. In the microwave part of the spectrum there are two main systems, *passive* microwave recorders and *active* microwave or radar, of which **side-looking airborne radar** (SLAR) is most used in remote sensing of the land surface.

The various sensor systems are discussed in Reeves (1975) and, with the exception of photographic ones, in a series of ESRO (now European Space Agency) publications (Higham *et al.* 1974, Ohlsson *et al.* 1972, Grant *et al.* 1973). Table 2.1 summarizes basic sensor and platform characteristics.

Sensors	Platforms	
	Low altitude (ground and aircraft)	High altitude (space, mainly satellites)
	Time of flight and area covered under control of operator (except free balloons). Cost borne by user.	Extensive and repetitive cover under constant illumination conditions. Orbit times fixed. Data often obtainable at less than cost.
Photographic (visible and near-infrared only)	High spatial resolution and geometric fidelity. Varied illumination conditions may be encountered. Difficult control of processing and calibration. Digitization requires separate process.	Civil uses in manned satellite missions and rockets only. Synoptic cover, high geometric fidelity.
Linescan (visible and infrared)	Usable over a wide wavelength range, with multiple registered channels possible. Lower spatial resolution and geometric fidelity than photography, better spectral resolution. Costly for aircraft use, with complex processing.	Suitable for unmanned satellites. Data recorded in digital or analogue form, can be telemetered to earth. Synoptic cover and constant orbits make geometric correction possible. Users may be able to obtain data at low cost.
Side looking radar (active microwave)	All weather system, can rapidly obtain cover of areas where photography impossible due to cloud. Limited availability of commercial systems – costly. Geometric and interpretation problems.	High power and weight requirements for satellite use. Only civil use so far in Seasat A. Spatial resolution is independent of range, so high resolution possible, but high data rates to match.
Passive microwave radiometer	Very low quantities of emitted energy, so antennae are large and complex, and data generally of lower spatial resolution. Interpretation principles not so well known.	Remarks re aircraft use also apply. Used so far in 'weather' satellites with spatial resolutions of tens of kilometres.

2.3.1 Photographic systems

Perhaps the best known and most commonly used remote sensing systems are photographic, using cameras and film. The film records the energy reaching it at exposure time in the visible and, in some cases, the near-infrared parts of the spectrum. Film may be either monochrome (black and white) or colour, the latter having three layers of emulsion sensitive to different parts of the spectrum, as opposed to the one layer in monochrome film. The spectral range of sensitivity of a film–lens system may range from the violet end of the visible spectrum at about 0.4 μm wavelength to about 0.9 μm in the near-infrared part of the spectrum. Both monochrome and colour films are available for the visible wavelengths only (respectively termed panchromatic and true colour) and for part of the visible and near-infrared range (respectively monochrome infrared and false colour infrared).

19

False colour infrared film, when processed, displays an image in unnatural colour to accommodate display of the infrared wavelength which has no visible colour (Table 2.2 and Plate 1). The range of any individual system will be restricted by the particular film characteristics and also perhaps by the use of filters.

Table 2.2 Colour renditions on colour films.

Film	Spectral region			
type	blue	green	red	near-infrared
True colour (e.g. Kodak Ektachrome 8442)	blue	green	red	
False colour infrared (e.g. Kodak Ektachrome 8443 with Wratten 12 filter)		blue	green	red

Filters may be used to make the filter–lens–film system sensitive to only a narrow waveband within the total range of the film. Problems may then result from the longer exposures necessary, so that compromises are often needed. Several such narrow-band film exposures may be made simultaneously in different bands of the total spectral range available. They are achieved either by using several separate cameras with parallel optical axes, or by a special single camera with multiple lenses illuminating different parts of the same film. Such multiple simultaneous narrow-waveband exposures are referred to either as multiband or multispectral photography (Wenderoth & Yost 1974). Examples of multispectral imagery, albeit obtained with a linescan rather than a photographic system, are shown in Figure 2.4.

2.3.2 *Linescan and related systems*

An alternative method of collecting data is a linescan system. In this system a mirror is rotated or oscillated about an axis parallel to the direction of movement of the aircraft or satellite platform (Fig. 2.2) (Higham *et al.* 1974). The mirror reflects the radiation received on to a detector, and the effect of the oscillation or rotation is to record data along a scan line perpendicular to the flight direction. The forward movement in flight or orbit of the platform ensures that scan lines are recorded successively further along the path of the platform, giving a second dimension to linescan data, and enabling images to be constructed.

As well as being used in the visible and near-infrared parts of the spectrum where photographic systems can also be used, linescan systems can, with suitable detectors, be used for any waveband as far as the microwave. In addition to visible

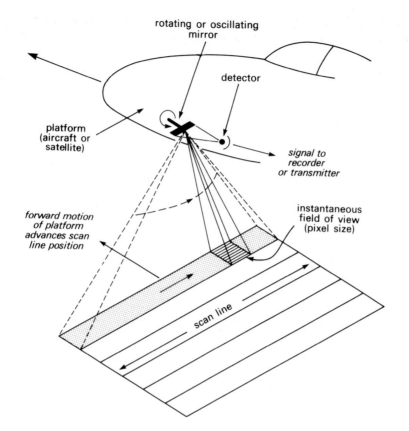

Figure 2.2 Principles of a linescan system.

and near-infrared, they are particularly used for narrow bands within the infrared having special properties, such as water absorption bands and, more commonly, for the detection of thermal radiation. Indeed, the term **infrared linescan** (IRLS) usually refers to a system for detecting thermal infrared radiation (the thermal radiation emitted from the Earth, proportional to surface temperatures) and not reflected near-infrared. An example of day and night thermal imagery is given in Figure 2.3.

Several types of scanner are available, frequently referred to as **multispectral scanners** (MSS), which record simultaneously a number of spectral channels, in some cases ten or more, in narrow wavelength bands through the visible, near-infrared and thermal parts of the spectrum. An example of three bands of multispectral scanner data obtained from an aircraft is given in Figure 2.4.

Perhaps the best known and most used MSS data are those provided by the Landsat satellite system. These data have been available since 1972, principally as MSS data, either in photographic image form or as digital computer compatible tapes (CCTs). The Landsat satellites are in a Sun-synchronous orbit, repeating a pass along a given track every eighteenth day, crossing the equator at approximately 0930 local time. Two satellites at a time have been operational for a considerable part of the time since 1975, giving either a nine-and-nine day interval, or six and twelve, between the satellites. A single scene covers an area of about 185 km × 185 km, with a nominal resolution provided by individual picture elements (pixels) of 79 m × 79 m. Data have been made freely available at reproduction costs or little more by NASA through their EROS Data Center at Sioux Falls, South Dakota. More

21

Figure 2.3 Thermal imagery
of an area near Reading,
England, obtained in the 8–14
μm waveband by a Daedalus
multispectral scanner flown in
an aircraft on 13 September
1977. Image A was obtained
at 06.00 hours, at minimum
temperatures, with a range
from black to 0° C to white at
18° C. Image B was obtained
at 14.30 hours, at maximum
temperatures, in the range 8° C
(black) to 44° C (white). The
swath is approximately 1·6 km
wide and 1·8 km long.
Thermal contrasts are
apparent notably of water
relative to land. The same area
is depicted in a false colour
infrared photograph in Plate
1, which also shows some of
the principal cover types, and
in Figure 2.4, depicting three
visible and reflective near
infrared wavebands, obtained
simultaneously with the 14.30
image above. Note the
geometric distortions in the
geometry of the two images,
common in aircraft
multispectral images.

recently Landsat data have also been available from a number of receiving stations around the world outside the USA, for example, in Canada, Brazil and Italy.

MSS data have been recorded in four separate spectral bands (channels): channel 4, 0·5–0·6 μm, the green band of the visible; channel 5, 0·6–0·7 μm, the orange–red band of the visible; channel 6, 0·7–0·8 μm, on the red–infrared boundary; and channel 7, 0·8–1·1 μm, in the near-infrared. An example of images from these four bands, recorded simultaneously, is given in Figure 2.5 with a 'colour composite' of three of these bands given in Plate 3. The best single source of information for the Landsat satellites is the *Landsat data user's handbook* (NASA 1976), and the revised edition (US GS 1979).

The linescan systems so far described use a rotating or oscillating mirror. Another system capable of obtaining a similar type of data is the **multilinear array** (MLA) detector. This is a research development at present, but is likely to be used operationally in the near future. As its name implies, it consists of an array of detectors arranged so that signals from a whole scan line can be detected simultaneously. The system has the advantage that there are no mechanical parts with potential to cause problems, but the detectors must all be radiometrically identical and must be pointed in exactly the right direction.

2.3.3 *Photographic and linescan systems compared*

A photographic system flown in a comparable platform will give a higher quality image insofar as it has better spatial resolution and geometric fidelity than that from a linescan system, but nevertheless the latter has advantages in some circumstances. It must be re-emphasised immediately that, for any wavelength beyond approximately 0·9 μm, a photographic system cannot be used; only a linescan with a suitable detector is possible.

The linescan detector produces a voltage varying according to the intensity of the radiation in the appropriate waveband falling on it; this voltage is amplified and can

Figure 2.4 Three bands of multispectral imagery obtained in narrow wavebands in the visible and reflective infrared parts of the spectrum, by a Daedalus scanner mounted in an aircraft at 14.30 hours on 13 September 1977, simultaneously with the false colour infrared photograph of Plate 1, and one of the thermal images in Figure 2.3. Plate 1 also shows some of the principal cover types. The wave bands are: (a) 0·50–0·55 μm (yellow/green, visible); (b) 0·60–0·65 μm (orange/red, visible); (c) 0·80–0·90 μm (near infrared). These and the images in Figure 2.3 and Plate 1 were obtained during the Joint Flight Experiment, UK, 1977, as part of the activities of the Tellus Project. The flight was financed and coordinated by the Commission of the European Communities, Joint Research Centre, Ispra, Italy.

(a)

(b)

(c)

(d)

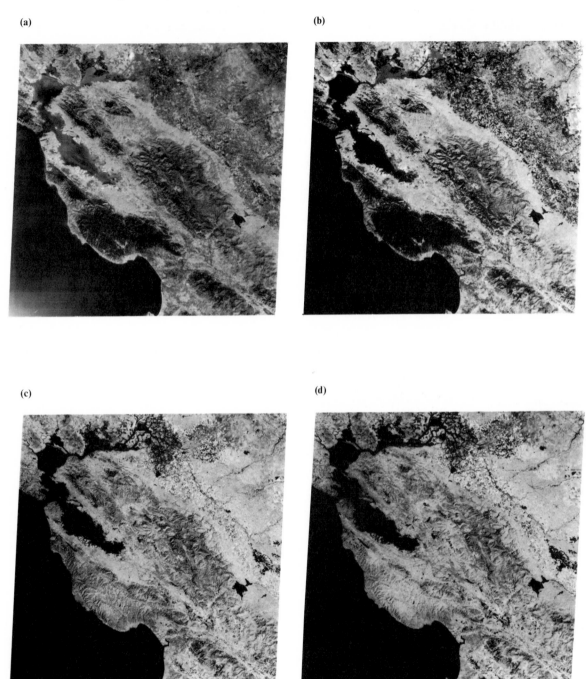

WI22-301
WI22-001
WI21-301
26OCT72 C N37-29/WI21-44 N N37-27/WI21-38 MSS   6   D SUN EL41 AZ146 190-1046-G-1-N-D-2L NASA ERTS E-1075-18173-6 02

24

be recorded as a signal on magnetic tape. This tape record can be processed to be computer compatible, either analogue or digital, whereas a photographic film requires scanning and digitizing before a CCT can be produced. There are further advantages for the voltage signal of linescan systems; firstly, if the detector is also exposed at regular intervals during use to radiation sources of known intensity, then the signal can be calibrated. With a photograph, absolute calibration is not feasible, and even approximate calibration is only possible if exposure and processing of the film is such that the grey scale lies entirely on the linear portion of the gamma curve of the film (e.g. Slater 1975, p. 262).

Another advantage of the linescan's voltage signal is that it can be transmitted directly to a receiving station at the Earth's surface with little loss of quality, whereas photographic film has to be physically recovered if full use is to be made of its characteristics (but note the remarks about return beam vidicon cameras below). With aircraft platforms this consideration is unimportant but for space platforms film is only recovered from manned missions or reportedly from specialized military satellites. For the usual unmanned programmes such as Landsat, data have to be transmitted to the ground by telemetry.

A sensor which combines some features of both photographic and linescan systems is the return beam vidicon (RBV) camera. It acts as a camera in that an image is obtained by a short exposure through a lens. Data are then stored for a short period, sufficient to read from the image plane and record or telemeter the data to Earth, rather similarly to the principle of a television camera. The image plane

---

Figure 2.5 Landsat-1 multispectral scanner (MSS) image of San Francisco and the Bay area, California: (a) channel 4 0·5–0·6 μm (yellow/green); (b) channel 5 0·6–0·7 μm (orange/red); (c) channel 6 0·7–0·8 μm (near infrared); (d) channel 7 0·8–1·1 μm (near infrared). Note the differences and similarities between the four images. Channels 6 and 7 are highly correlated, though there are differences for example in the degree of absorption in the inner estuaries in the central north of the image. The redwood forests adjacent to the coast south of the Bay are represented by a dark tone on the channel 5 image indicating high absorption in this channel, in contrast to channels 6 and 7. Close inspection reveals many other differences between channels 5 and 7 in the cultivated Sacramento Valley in the north east of the image. Note that channel 4 has an overall hazy appearance since it suffers most from atmospheric scattering. Channel 4 also reveals most detail in water areas, whereas in channel 7 there is almost complete absorption (see Fig. 2.8). Area depicted: 185 × 185 km. Plate 3 is a colour composite of channels 4, 5 and 7. NASA image E-1075-18173.

Below all Landsat images a detailed technical description is provided. Explanation of the caption of the above images is provided below, which is representative of Landsat photographic products created before 18 February 1977. A representative example of the captions after this date is provided in Fig. 10.4a.
06 OCT72 date of imaging; **C N37-29/W121-44** latitude and longitude of the point vertically below the satellite when imaging the central point of the image; **MSS 6 D** sensor, spectral band number, direct transmission to Earth (**R** if played back from the satellite recorder); **SUN EL 41 AZ 146** elevation and azimuth of sun to nearest degree; **190-1046-G-1** spacecraft heading in degrees – revolutions of Earth since launch – ground receiving station identifier, namely Goldstone; **1-N-D-2L** full size image – normal (**N**) or abnormal (**A**) processing – data used to compute image centre, predicted (**P**) or definitive (**D**) – mode of transmission to Earth: linear, **1**, or compressed, **2**, – high (**H**) or low (**L**) gain of sensors 4 and 5; **NASA-ERTS** agency and name of project; **E-1075-18173-5** scene identification number (used in ordering imagery), defined as follows:
E project letter (viz ERTS)
1 satellite number (1, 2 or 3)
075 day number since launch
18 hour relative to day number
17 minute relative to day number
3 tens of seconds relative to day number
6 spectral band number
Further details can be found in NASA (1976) and USGS (1979).

25

can then be ready for the next exposure: no replacement of film is necessary and the data are acquired in a suitable form for magnetic tape output. R B V cameras have been used on the Landsat satellites, and the latest of the series, Landsat 3, carries a R B V system capable of providing Earth surface imagery with a pixel size about 24 m × 24 m (Fig. 6.8).

Since linescan and R B V camera data can be converted to CCTs with little loss of quality, output from these systems may be preferred where computer assisted analysis is proposed. If visual analysis is to be used, then photographic products may be preferred. Of course, images can be produced from linescan data, and digital data from photographic images, but both these conversions involve some reduction in quality of the original data.

2.3.4 *Microwave systems*

The remaining part of the spectrum used for Earth resources studies is the microwave range from about 1 mm to 1 m in wavelength. Microwave radiation is emitted from the Earth's surface, but in very small quantities in the long 'tail' of the Earth emission spectrum (see Fig. 2.1) (Ohlsson *et al.* 1972). Detection and recording of these minute quantities of radiation present great technical problems, and the data collected are of lower resolution than those in shorter wavelengths. Research work has been carried out using aircraft platforms, and microwave data

Figure 2.6 Principles of side-looking airborne radar (SLAR).

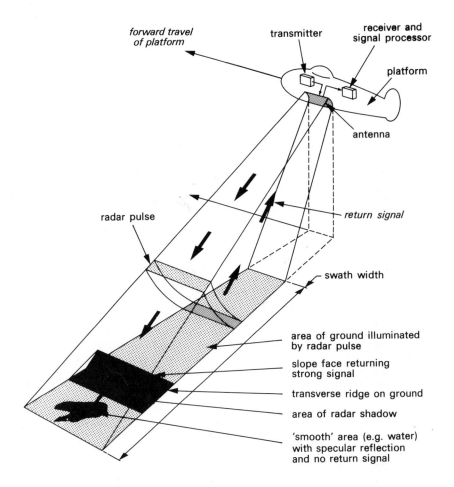

forward travel of platform

transmitter

receiver and signal processor

platform

antenna

radar pulse

return signal

swath width

area of ground illuminated by radar pulse

slope face returning strong signal

transverse ridge on ground

area of radar shadow

'smooth' area (e.g. water) with specular reflection and no return signal have been recorded on a number of atmospheric observation satellites, with resolutions of several kilometres.

Systems for recording the microwave radiation emitted naturally from the Earth are known as *passive* microwave systems, as distinct from systems which sense their own emitted radiation after its reflection at the Earth's surface: the latter systems are known as *active* microwave systems, or more commonly as *radars*. In a radar system, pulses of microwave radiation are ('actively') produced by a transmitter carried on the platform at the same position as the receiver or sensor. The level of energy of these pulses is locally far above the emitted or passive microwave energy. The radar receiver collects and records that part of the 'active' pulsed energy returned to the sensor. An approximate analogy may be made with a flash photograph taken in conditions of near darkness, in which the ambient radiation is insufficient to be recorded at the sensor (the film), so artificial energy is supplied (by the flash).

Radar systems have been developed in many forms, but the one most generally used in Earth resources surveys is SLAR (Fig. 2.6), (Grant *et al.* 1973, Moore 1976). In the simplest form of this system, radar pulses are generated in a plane perpendicular to the direction of travel of the platform. The antenna is designed so that the pulses, while being comparatively wide in vertical extent, are very short in duration and narrow in the flight direction. Thus at any given instant only a very small area of ground, of perhaps tens of metres or less in size, is illuminated by the radar pulse. According to the nature of the terrain, a varying proportion of this energy is reflected back to the platform. As the pulse travels outwards from the platform it will illuminate different strips of ground successively, and the time of travel of the returned portion of the energy will increase with increasing distance of the illuminated piece of ground (Fig. 2.6). Thus, using the varying time delay, the return signal can be recorded along a line perpendicular to the direction of travel of the aircraft, analogous to a single scan line of a linescan system. In the same way, the forward travel of the platform enables successive strips of ground in the direction of travel to be illuminated, and the data recorded on magnetic tape or on film can be reconstituted as a radar image, of which Figure 2.7 gives an example.

Figure 2.7 Side-looking radar image of part of the Peak District, Derbyshire, UK (8·6 mm wavelength). Illumination is from the upper (eastern) edge of the image. The steeper slopes facing westward are not illuminated by the system's radiation (see Fig. 2.6) and hence are in total shadow. The latter slopes are escarpment slopes formed in resistant gritstones, which dip towards the east. Area depicted approximately 8 × 4·5 km (UK Crown Copyright, Ministry of Defence).

The above description refers to real aperture or 'brute force' systems. More complex synthetic aperture systems are available, in which a Doppler effect is used to produce images of higher resolution size for any given antenna size.

These radar images are at first sight broadly similar to a photographic or line-scan image, but their properties, both geometrical and interpretative, differ. The proportion of energy returned varies with the slope and roughness of the ground surface and with its complex dielectric constant (Moore 1976). To the skilled interpreter, radar images can often reveal information additional to that provided by other images, and radar data are increasingly being used in conjunction with other data.

2.4 Types of platforms and multilevel sampling

In the above discussions of sensors and data for various parts of the spectrum, frequent mention has been made of the variety of platforms possible in terms of aircraft and spacecraft. A more strict separation should perhaps be made between low-altitude and high-altitude platforms. There is a continuum of platform altitudes, between the camera or other sensor, hand-held or mounted on a structure attached to the ground, through such low-altitude platforms as kites, tethered balloons, remotely controlled pilotless aircraft, autogiros, helicopters and low flying fixed wing aircraft, to the high flying fixed wing aircraft at around twenty kilometres, and unmanned free balloons which may fly at up to thirty or forty kilometres. All these may be termed low-altitude platforms, in contrast to rockets and, much more commonly, satellites, which carry sensors at altitudes between two or three hundred kilometres and the 36 000 km required for a geostationary orbit (although rockets may operate below this lower limit). There is a considerable break in operating altitudes between the two groups, with the latter group being termed high-altitude platforms.

There are obvious broad differences between the two groups of platforms. The times of flight of low-altitude platforms and rockets are under the control of the operator, so that the most suitable time may be chosen, or the flight may be delayed if conditions are unsuitable for weather or technical reasons. Such a delay will, of course, incur a cost, which may be very considerable in the case of a conventional aircraft and crew. Costs increase almost directly proportionally with the number of missions flown, so that repetitive cover by low-altitude platforms may not be undertaken lightly. The absolute cost of a low-altitude mission or group of missions will almost certainly be less than the cost of launching a satellite. Relatively, however, when one considers the fact that a satellite, once launched, will continue to function in its orbit and give repetitive cover, provided there is no technical breakdown, without further satellite costs, a satellite can become an attractive proposition. It must be remembered, of course, that ground receiving station costs are continuous for the life of the satellite. Satellite data are extensively used, but a major point here is that the cost of the satellite is, in most cases, not borne by the user, and data are made available very cheaply.

In the circumstances, the extensive and repetitive cover provided by satellites is very attractive. There may be disadvantages to this, however, for a satellite orbit is fixed in time and space, at least in the short run, so that Earth surface data may be lost for successive passes due to cloud cover, for example. A single Landsat satellite

28

Table 2.3 Earth resources satellites.

Manned satellites

Mercury, Gemini and Apollo missions	Early manned space photography, mostly hand-held and 'opportunity' oblique Hasselblad 70 mm photography.
Skylab	Launched 1973: mixed missions, cover *c.* 50°N to 50°S. Carried a six band multispectral camera, a high resolution camera (Earth Terrain Camera), a 13 channel scanner covering visible and infrared regions, a microwave radiometer, and a radar altimeter/scatterometer radiometer. Limited data available (*Skylab Earth Resources Data Catalog*, NASA, 1974).
Spacelab	Planned as re-usable manned vehicle, to be launched by the Space Shuttle, late 1980 onwards. Will carry a variety of sensors: early planned missions include a metric camera and microwave experiments. Experimental only.

Unmanned satellites

Landsat satellites	Landsat 1: July 1972–Jan. 1978. 4 channel MSS (0·5–0·6 μm, 0·6–0·7 μm; 0·7–0·8 μm; 0·8–1·1 μm), and 3 channel RBV (0·475–0·575 μm; 0·580–0·680 μm; 0·698–0·830 μm) (Section 2.3.2) *c.* 79 m resolution on all systems. Little RBV data collected. Data available as images and CCTs.
	Landsat 2: Jan. 1975 continuing as at August 1980. Sensor system as Landsat 1.
	Landsat 3: March 1978, continuing. MSS as Landsats I and 2 except extra thermal channel, *c.* 240 m resolution, failed July 1978. RBV: single channel (0·505–750 μm) *c.* 24 m resolution. (see *Landsat data users handbook*, NASA, 1976 and USGS, 1979).
	Landsat D: 'Follow-on', planned for launch 1982. To carry 5 band MSS compatible with present system, and a new 7 channel Thematic Mapper with 30 m resolution (0·45–0·52 μm; 0·52–0·60 μm; 0·63–0·69 μm; 0·76–0·90 μm; 1·55–1·75 μm; 2·08–2·35 μm; 10·40–12·50 μm) except for the last with 120 m resolution.[1]
Heat capacity mapping mission (HCMM) (also referred to as applications explorer missions A(AEM-A))	Launched April 1978. Some data available to date. Designed for observation of thermal emissions in waveband 10·5–12·5 μm, also visible and near-infrared in waveband 0·5–1·1 μm. Orbit designed to give mid-latitude cover at times of maximum and minimum temperatures in a 24 hour period, any point covered approximately every 8th day. Resolution *c.* 700 m, swath width *c.* 700 km. Experimental, for investigation of determination of terrain data, particularly soil moisture, from day–night differences.
Seasat-1 (ocean dynamics measuring satellite)	Launched April 1978, no longer operational. Limited data available of land areas. Carried a radar altimeter, radar scatterometer, synthetic aperture imaging radar, a scanning multifrequency passive microwave radiometer and a visible/infrared scanning radiometer.

Table 2.3 – *continued.*

Atmospheric observation satellites

The general objective of these satellites is to provide low resolution cover of wide areas at frequent intervals. Some may provide Earth surface images or data, generally not suitable for Earth resources surveys, but which may be of limited use, and which warrant investigation in any given case.

A large number of these satellites has been launched since 1960, including such series as Tiros, Cosmos, Essa, ATS, Meteor, ITOS, NOAA, SMS/GOES, Nimbus and Meteosat. A summary of the earlier satellites and their data was given by Barrett (1975).

Currently operational satellite systems include:

NOAA series	1972 onwards, polar orbiting satellites, global coverage twice daily. Visible waveband (0·5–0·7 μm) 3·6 km nominal resolution, and thermal infrared (10–12 μm) at 7·2 km. Very high resolution radiometers (VHRR) provide cover of limited areas in the same bands at *c.* 0·9 km resolution. Being replaced by TIROS-N and subsequent satellites.
SMS/GOES satellites	1975 onwards. Geosynchronous satellites at *c.* 35 800 km orbit above equator. Visible and infrared sensors, imagery at *c.* 30 min frequency. GOES-1 at approx. 140° W, GOES-2 at 75° W approx.
Meteosat	Launched November 1977. Geosynchronous orbit, 0° longitude as part of global atmospheric research programme (GARP), to operate with GOES satellites. Details similar, resolution *c.* 2·5 km in visible *c.* 5 km in infrared. Best imagery is of Africa.
Nimbus series	Experimental programmes for development of atmospheric research. Latest is Nimbus G launched 1978, carrying a package of sensors for investigation of Earth radiation budget and ocean surface measurements.

Planned developments

Satellites are planned by many countries and agencies. Among those particularly worthy of mention are the European Space Agency's (ESA) SARSAT (synthetic aperture radar satellite) and PAMIRASAT (passive microwave radiometer satellite). A French satellite, SPOT (satellite probatoire pour l'observation terrestre) with a scanner resolution of 10–20 m, is planned for launch in 1983/4. SPOT wavelengths, 0·50–0·59 μm; 0·61–0·69 μm; 0·79–0·90 μm. A stereosat to give stereoscopic imagery has been proposed by the Jet Propulsion Laboratory (JPL), California Institute of Technology with a 15 m resolution and a wide spectral band in the visible wavelengths.

---

[1] At the time of going to press, problems in construction of the Thematic Mapper may mean that Landsat D only has the MSS on board and the Thematic Mapper is installed in Landsat D′.

passes over any given point at a fixed time on every eighteenth day and given a low probability of cloud-free conditions, it is easy to see that good surface data will be obtained rarely. This is a frequent problem in mid-latitude and equatorial areas.

The area covered by a single satellite scene is generally very large in comparison with low-altitude data, a single Landsat frame covering approximately 185 km × 185 km, about 34 000 km², for example. This is generally at the cost of reduced spatial resolution, making the detection and recognition of small features difficult or impossible. Using the Landsat MSS system as an example again, a single picture element (pixel) represents a ground area approximately 79 m × 79 m, while objects of fractions of metres dimension of adequate contrast can be recognized on aircraft film. The extensive areas covered by satellite data may, however, make it easier to incorporate them into any data bank, and there is the added advantage that a single scene will be acquired in a few seconds, under solar illumination conditions that are virtually uniform over the whole scene (at least at the top of the atmosphere).

Summary details of the principal satellite systems whose data may be used in Earth resources surveys are given in Table 2.3.

A solution which may be as nearly ideal as can be obtained, may frequently be provided by combining remotely sensed data from different types of platforms with ground data in a *multilevel sampling* scheme. For example, at the same time as a scheduled satellite pass over an area, aircraft underflights of test areas, carrying suitable sensors, should be arranged, and teams should be collecting simultaneous data on the ground for the same test areas. Combinations of these data at different resolutions from different altitudes will almost certainly provide more useful data than the use of any single source. The use of such schemes is recommended wherever possible as an essential part of an integrated approach, even if the data cannot be collected at the same time.

2.5 Uses of remotely sensed data

2.5.1 *Uses of visible and near-infrared data*

By far the greatest amount of remotely sensed data is available in this part of the spectrum, and more use is made of these data than of any other. It is preferable to consider these two parts of the spectrum together, since their radiation has similar physical properties and both may be recorded by photographic and linescan systems. Data are available from aircraft and from satellite platforms. A major obstacle to data acquisition is that direct radiation in these parts of the spectrum cannot pass through cloud cover. Thus much satellite imagery of the Earth is of cloud tops rather than the Earth's surface and aircraft photography is frequently hindered or prevented by cloud conditions.

The data forms most commonly used for Earth resources surveys are aircraft photography and Landsat MSS data, with other sources such as aircraft borne linescan systems and other satellites a long way behind at present for this use.

Visible and near-infrared data have been used for a multitude of purposes. Aircraft photography has been for several decades and still is the basic data for photogrammetric, topographic mapping. This is now a well developed independent science and as such will not here receive more than this passing mention (see e.g. Kilford 1963, Wolf 1974, or Thompson 1966, for discussions of photogrammetry at various technical levels).

A more recent development, since the development of linescan systems and multiband photography, is the separation of land cover types and other land characteristics by analysis of the spectral characteristics of remotely sensed data (e.g. Swain & Davis 1978).

Different land surfaces reflect different proportions of the solar radiation falling upon them in different spectral bands (e.g. Fitzgerald 1974). This is illustrated in a simplified style in Figure 2.8. The diagram illustrates on a broad scale the differences between snow and ice (strongly reflecting at all wavelengths), water (absorbing at all

31

wavelengths), bare soil (steadily increasing reflection as wavelength increases) and vegetation (with a weak peak of reflection in the green, strong reflection in the near-infrared, and absorption in the blue and the red). It also illustrates on a more subtle scale the differences between new snow and old dirty snow and ice (lower reflection from the latter), between deciduous and evergreen vegetation (low reflection in the near-infrared from evergreens), between dry and moist soil (reduced reflection as moisture content increases) and between clear and turbid water (increasing reflection in the shorter wavelengths with increasing turbidity).

The diagram, although much simplified, indicates the nature of distinctions that may be made from image *tones* in different spectral bands, and thus the value of the multispectral approach. A hierarchical system of classification may therefore be used. For example, water can usually be separated from all other areas (except cloud shadows) on a Landsat scene by the low values in channel 7. The second stage may then be a distinction between clear and turbid water from values in channel 4. Green vegetated areas may usually be distinguished from others by determining the ratio value channel 5/channel 7 for all pixels: it tends to be much lower for vegetated areas. Differentiation within the vegetated areas is then more difficult, requiring more subtle analysis for such tasks as separation of vegetation types, stress, phenological stage and so on.

Such differences in reflectance of land surfaces led to the coining of the term 'spectral signature', but the implied metaphor has proven inexact since in practice the reflectance of a given land surface often changes through time and may often be

Figure 2.8 Reflectance values of different land cover types at different wavelengths in the visible and near infrared.

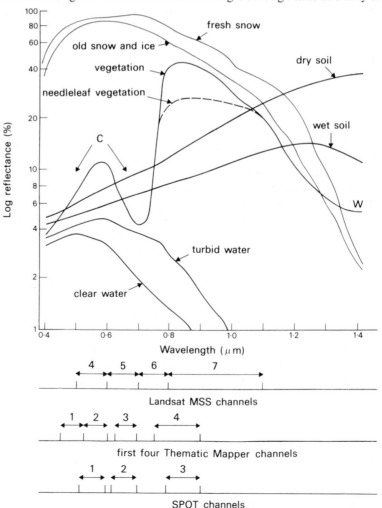

far from unique. Hence, in recent years, the term has tended to be eschewed in favour of **spectral response**.

It must be stressed in conclusion that the bulk of all work using remotely sensed data uses these spectral areas of the visible and near-infrared. Such work is by now more or less routine and operational, whereas much use of the other spectral areas about to be considered is in the research and development stage.

2.5.2 *Uses of thermal infrared data*

Thermal radiation is emitted from the Earth's surface, in contrast to the reflected solar radiation that has been considered in the previous section. All natural surfaces are nearly perfect emitters of energy in the thermal band, while in the waveband 8–14 μm, which includes the peak energy emission from the Earth, a cloudless atmosphere is nearly transparent to thermal radiation. In the observed range of Earth surface temperatures, commonly from about 270 to 330 K, the radiation received at a remote sensor at aircraft altitudes may be assumed as a first approximation to increase linearly with the surface temperature.

Recording of thermal emitted radiation on a spatial basis requires a linescan system. Equipment with a single thermal channel, usually within the 8–14 μm 'window' is most commonly available. These are for mounting in aircraft, and are commonly of military parentage.

The Earth resources surveyor needs to interpret subtle thermal effects such as the difference between dry and moist soil and between transpiring and other vegetation. A single set of thermal data for one time can be useful in this connection, if the interpreter makes use of his estimates of the heat balance and of the thermal properties of materials – a water body is conservative, dry soil will have a large diurnal range of temperature, and so on. Thermal data are best acquired at times of high temperature contrast, shortly after mid-day or before dawn. Mid-morning or evening flights are likely to show much lower temperature contrasts. Unexpected effects may result from such phenomena as shade from the Sun, protection from the wind, cold air drainage or shallow mist formation. Springs and seepages may be detected at their outflow into a water body because of the contrasting water temperatures. This may provide useful indicators of water sources and geological structure. Thermal radiation is of course independent of daylight, but will not penetrate cloud, fog, mist or precipitation. Remotely sensed thermal data for a single time can be useful in the ways described above, but if two sets of data are available within a 24 hour period, at times of maximum and minimum temperatures, they can yield more information, provided the day and night values for each point can be registered to yield values for temperature differences. Given suitable ground information, thermal inertias can be computed and extrapolations made from test areas. Thermal inertia data have been used in arid, vegetation free areas to detect differences in rock and soil types (Kahle *et al.* 1975). Where moisture is present, thermal inertia values can provide an indication of moisture content of bare soil and rock. The presence of vegetation complicates interpretation greatly.

Remotely sensed thermal data have been available from aircraft platforms, and from atmospheric observation satellites such as the NOAA series at resolutions of kilometres or slightly less. 1978 saw the launch of two satellites designed to produce thermal data for use in Earth resource studies. The first, Landsat 3, produced little thermal data before the system failed. The second was the heat capacity mapping mission (HCMM) satellite, with an orbit designed to provide 12 hour cover for mid-latitude areas at approximately the times of maximum and minimum temperature. Resolution of the thermal data is approximately 500–700 m, and swath width is 700 km (Fig. 4.2). In addition to the thermal data, another data channel records the radiance value for reflected solar radiation in the 0·5–1·1 μm band, thus enabling short-wave albedo to be determined.

2.5.3 *Uses of microwave data*

The uses of these data to the Earth scientist depend on two distinct properties of microwave radiation. The first of these is that, because of its longer wavelengths, it penetrates cloud cover without appreciable loss. (However, it will not penetrate areas of heavy rain, which has the advantage for the meteorologist that areas where rain is falling can be distinguished from cloudy areas where no rain is falling.) This property gives rise to the claims for microwave systems as 'all-weather' equipment.

The second distinctive property of emitted microwave radiation is that different Earth surface materials have very different microwave emissivities (for example, the microwave emissivity of a soil varies with changing moisture content). The emissivity of a substance varies between 0 and 1, and is the proportion of energy of a given wavelength emitted by the substance at a given temperature compared with that emitted by a perfect radiator at the same wavelength and temperature. Thus the emissivity, ε, of a perfect radiator, or 'black body', is 1. These varying microwave emissivities are in contrast with the situation in the thermal waveband where, as already noted, nearly all natural surfaces are nearly perfect radiators. Using microwave emitted radiation, therefore, surfaces of nearly identical temperatures can be readily distinguished, using the concept of microwave

Figure 2.9 Synthetic aperture radar (SAR) image of Washington D.C. obtained from Seasat 1 (see Table 2.3), taken from 800 kms in the 23·5 cm microwave band. Ground resolution is approximately 25 m. The Potomac River at the bottom right (southwest) of the image and the Anacostia River in the upper part (east) of the image appear dark because of the almost perfect specular reflection from the water surfaces, so that little or none of the signal returns to the sensor (Fig. 2.6). These rivers bound the downtown part of Washington DC, where a large number of very bright return signals can be seen. This is typical of urban areas due to the high frequency of reflective surfaces at right angles to each other (corner reflectors) returning a high proportion of the incident signal to the sensor. The broad dark zone which extends from the centre of the image to the Potomac River is the grass-covered Mall which leads up to the Capitol building. The five-sided shape of the Pentagon can be readily picked out in the bottom right of the image. Area depicted: 11 × 11 km (NASA image).

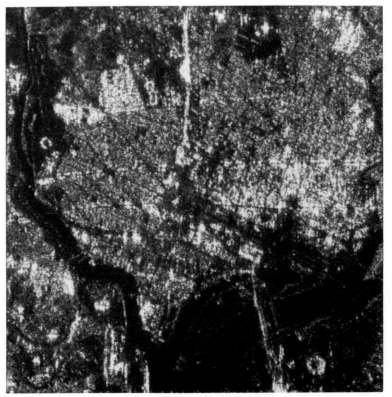

'brightness temperature' (BT) as opposed to 'thermal' temperature. Since, for example, the microwave emissivities of ice and water differ considerably, a passive microwave image from a satellite permits the determination, through a cloud layer, of the ice–water boundary in the oceans.

It may be concluded, however, that, primarily because of low resolution, emitted (or passive) microwave data are likely to remain of little use in the immediately foreseeable future to Earth resource surveyors. However, research is in progress, although the use of passive microwave sensors in aircraft is rare.

The same conclusions do not apply to 'active' microwave, or side-looking radar. The 'active' microwave radar energy will have the same properties as 'passive' emitted microwave energy, in that it is unaffected by clouds. Radar surveys can thus be carried out by day or night in all weather conditions except heavy precipitation and thus, if the data can be presented in a suitable form, such surveys can be of great advantage in areas that are almost continually cloud covered, where conventional air photography may be impossible for very long periods. For these and other reasons, side-looking radar equipment has been flown extensively in aircraft over land areas in the last decade or so, and there are several commercial organizations able to provide side-looking radar surveys. Since mobilization and operational costs are high, a radar survey is only feasible for large areas if any cost-benefit justification is expected. Such surveys have been carried out, for example, for the Amazon basin area of Brazil and for Nicaragua, and a survey has been interpreted for the forested areas of Nigeria. Part or all of a number of other countries, including Peru and Indonesia, have also been flown for side-looking radar cover.

Such surveys and others have been used as a topographic base in areas for which no adequate maps are available, because continuous cloud cover had prevented aircraft photography. They have been interpreted for forest inventory, drainage net delineation and the interpretation of the relation between surface form and geological structure. Radar images obtained from aircraft are particularly well suited for this latter purpose: they have a superficial resemblance to low-angle Sun photography, with strong shadows showing structures well (Fig. 2.7).

Until recently, power and weight requirements have prevented an imaging radar being flown on a satellite for unclassified civil purposes. In 1978, Seasat-1 was launched, carrying a synthetic aperture radar system with a nominal resolution of 25 m, although the satellite had a short life of some three months (Fig. 2.9).

Radar theory shows that the resolution of a synthetic aperture system is independent of target range, so that a satellite radar system should be capable of very high resolution. However, this advantage has to be paid for by either a reduced swath width or an enormous data transmission rate, which would require a large bandwidth (Moore 1976).

2.6 Relating remotely sensed data to other geographic data bases

Our capability for acquisition of remotely sensed data far exceeds that for their processing and interpretation. A very high proportion of the data available has never and will never be fully used.

There is, of course, a variety of reasons for this situation which has been true throughout the development of remote sensing. The contrast in rates of data acquisition and processing has always been large. Photography acquired in a day's flying is sufficient to keep a photogrammetric team busy for months, while a Landsat scene consisting of about eight million pixels in each of four channels is

35

acquired in about 25 seconds, and may be the subject of weeks or months of analysis.

One of the outstanding problems in this situation is the difficulty found in physically integrating the remotely sensed data with the other available material. Most of these data, remotely sensed or otherwise, are spatial in character, and the total information system must be firmly fixed to a spatial base. In most countries this base must be a topographic map system, where maps are available to a sufficient standard. In some parts of the world the good topographic map base is not available, and a photomosaic or, more recently, a Landsat based photomap may provide the reference framework.

Whatever the system, knowledge of the location of spatial data is vital if good use is to be made of them, and thus remotely sensed data, to be part of a complete information system, must be registered to the local map base. This may be done by visual interpretation and manual transfer of interpreted information, or by an understanding of the geometrical properties of the remotely sensed data and suitable processing based on this understanding (e.g. Hardy 1978).

Examples of such processing include the preparation of orthophotos or orthophotomaps from stereoscopic air photographs. Such a process makes use of the stereoscopic properties of photographs to eliminate the distortions caused by these very properties. Aircraft-borne linescan systems are usually based on a constant velocity rotating mirror, sampled at a constant time interval, which produces its own peculiar distortions. These may be corrected either by digital resampling or by producing prints or corrected negatives by using suitably curved surfaces, depending on the nature of the data.

Satellite data such as those from Landsat usually require major corrections for scaling, rotation, and shear effects caused by Earth rotation beneath the satellite's suborbital path, with minor non-linear corrections resulting from a variety of other effects such as changes in the satellite's attitude and altitude. Covering larger areas in a more uniform fashion, these data are generally much more amenable to correction to a map base, by resampling methods for digital data. The resampled data can be either the original radiance values, available for later classification, or classified data. Digital data can be combined in the same system with digitized map data, and ground data (Section 3.7), enabling classification, comparison and correction to be made within the data bank, and forming a true geographical information system (Section 5.4).

References

Barrett, E. C. 1974. *Climatology from satellites*. London: Methuen.

Barrett, E. C. 1975. Environmental survey satellites and satellite data sources. *Geography* **60**, 31–9.

Farrow, J. B. 1975. *The influence of the atmosphere on remote sensing measurements*. European Space Agency, Neuilly-sur-Seine, France, 3 vols. ESA (ESRO) CR 353–5.

Fitzgerald, E. 1974. *Multispectral scanning systems and their potential application to earth-resources surveys: spectral properties of materials*. ESRO (now European Space Agency), Neuilly-sur-Seine, France. ESRO CR-232.

Fleagle, R. G. and J. A. Businger 1963. *An introduction to atmospheric physics*. London: Academic Press.

Grant, K. *et al.* 1973. *Side-looking radar systems and their potential application to earth resources surveys*. ESRO (now European Space Agency), Neuilly-sur-Seine, France, 6 vols, ESRO CR 136–41.

Hardy, J. R. 1978. Land cover studies using Landsat MSS digital data: cartographically
corrected lineprinter output. *Proc. 12th Int. Symp. on Remote Sensing of Environment, Ann Arbor, Michigan*, 1717–27.

Higham, A. D. *et al.* 1973–75. *Multispectral scanning systems and their potential application to earth resources surveys.* ESRO (now European Space Agency), Neuilly-sur-Seine, France, 7 vols, ESRO CR 231–7.

Kahle, A. B., A. R. Gillespie, A. F. H. Goetz and J. D. Addington 1975. Thermal inertia mapping. *Proc. 10th Int. Symp. on Remote Sensing of Environment, Ann Arbor, Michigan*, 985–94.

Kilford, W. K. 1963. *Elementary air survey*, new edn in press. London: Pitman.

Lintz, J. and D. S. Simonett 1976. *Remote sensing of environment.* Reading, Mass.: Addison-Wesley.

Monteith, J. L. 1973. *Principles of environmental physics.* London: Edward Arnold.

Moore, R. G. 1976. Active microwave systems. In *Remote sensing of environment*, J. Lintz and D. S. Simonett (eds), 234–90. Reading, Mass.: Addison-Wesley.

NASA 1974. *Skylab earth resources data catalog.* NASA, Goddard Space Flight Center.

NASA 1976. *Landsat data user's handbook.* NASA, Goddard Space Flight Center.

Ohlsson, E. *et al.* 1972–73. *Passive microwave radiometry and its potential application to earth resources surveys,* ESRO (now European Space Agency), Neuilly-sur-Seine, France, 6 vols, ESRO CR 71–5 and 116.

Reeves, R. G. (ed.) 1975. *Manual of remote sensing*, 2 vols. Virginia: American Society of Photogrammetry.

Sellers, W. D. 1965. *Physical climatology.* Chicago: University of Chicago Press.

Slater, P. N. 1975. Photographic sensors for remote sensing. In *Manual of remote sensing*, R. G. Reeves (ed.), 235–323. Virginia: American Society of Photogrammetry.

Swain, P. H. and S. M. Davis (eds) 1978. *Remote sensing – the quantitative approach.* New York: McGraw-Hill.

Thompson, M. M. (ed.) 1966. *Manual of photogrammetry*, 2 vols. Virginia: American Society of Photogrammetry.

USGS (United States Geological Survey) 1979. *Landsat data user's handbook*, revised edn. Arlington, Va.: US Geol. Survey.

Wenderoth, S. and E. Yost 1975. *Multispectral photography for Earth resources remote sensing.* New York: Information Center.

Wolf, P. R. 1974. *Elements of photogrammetry.* New York: McGraw-Hill.

3 Integrating ground data with remote sensing

Christopher O. Justice and John R. G. Townshend*

3.1 Introduction

In the past ground data collection was the pre-eminent source of information for land resources survey, but remote sensing imagery has reduced its importance for many types of survey. Ground data collection nevertheless remains an essential element even in surveys heavily dependent on remote sensing (Pettinger 1971).

The relative importance of ground data and remote sensing data in a survey depends upon several factors, including the survey's objectives, its scale, budgetary and time constraints, the region's accessibility, and the number of trained personnel; also of importance are the types of imagery used, and the effectiveness of processing and classification methods that are available. Even if all the above factors are known, there is as yet no well defined methodology for deciding on the proportional expenditure of money, time, and expertise on remote sensing and ground data collection. Controversy concerning the confidence to be placed on remote sensing data is long-standing. In the specific context of panchromatic air photographs as an aid for soil survey, Belcher (1948) stressed their importance and the relative unimportance of ground survey, a view challenged shortly afterwards by Hittle (1949) and subsequently by Buringh (1960) and Vink (1968). For example, Vink (1968) states that air photo-interpretation can decrease field observations carried out for plotting boundaries of soil-mapping units but is of no help in the identification of the soil units actually present.

The introduction of newer methods both of remote sensing and data analysis has rightly led to the belief that the intensity of ground survey can often be reduced. Unfortunately, in many remote sensing studies, it is clear that the field component is regarded as relatively insignificant. Explanation for this lies partly in the belief that ground data are easily obtainable and that archival surrogates are usually adequate. Such overconfidence in ground data and the ease with which they may be collected is highlighted by the bastard term 'ground truth', often used as a surrogate for 'ground data' or 'ground information'. This term is usually inaccurate and misleading since it implies the ground data are free from error.

Ground observations are subject to instrumental errors, operator errors, ground location errors, errors induced by observations not being synchronous with image collection, and errors arising from inadequate sampling design. Their minimization is clearly desirable for any ground survey but becomes especially important where remote sensing is the dominant data source, since the reliability of the collected ground data will strongly affect the validity of extrapolations to areas not visited in the field. This chapter is concerned with methods for ensuring that ground data are collected both efficiently and accurately for terrain resources survey. In keeping

*C. O. Justice was a National Research Council resident research associate at NASA/GSFC, Maryland, during the writing of this chapter.

with the main theme of this book, we will concentrate on the collection of ground data combined with remote sensing data though many points are relevant even when this is not the case. The phrase 'reference data' has usefully been suggested to include all data used for interpreting remote sensing imagery (Swain & Davis 1978), of which ground data is normally the most important component.

In an operational survey utilizing remote sensing imagery, the roles of ground data collection are diverse (Benson *et al.* 1971). Where ancillary data are limited, field work can help in the survey planning stage including improvements in the design of sensor configuration (Lee 1975, Tucker & Maxwell 1976). Ground data are also needed to train human or computer classifiers to extrapolate over a whole region or target area, using correlations between ground data and image properties (Section 4.9). Assessment of the viability of these extrapolations requires further ground data for the creation of testing sets. Selection of testing and training sets and the evaluation of the subsequent statistical analysis demand considerable care. Moreover, ground data can be used for adjusting areal estimates of terrain characteristics derived from remote sensing sources by applying weighting coefficients. Bauer (1977) summarizes one example of the method of making adjustments as follows:

$$P = (E^t)^{-1} \ddot{P}$$

where

P = vector of corrected proportions
$\ddot{P}$ = vector of proportions estimated from the remote sensing data
E^t = transpose of the matrix of training field classification, where its elements
e_{ij} are the proportion of samples of type i classified as type j.

Ground data are also acquired for research purposes usually in an intensive manner at a relatively small number of test sites. Such investigations are summarized by Lee (1975) and Lintz *et al.* (1976) and although largely outside the scope of this book are vital for a better understanding of the information potentially available from remotely sensed data.

As a final preliminary we note that, statistically speaking, *sampling* is a method of selection from a larger population, carried out in order to reduce the time and cost in examining the whole population. The resultant individuals that are sampled are known as *sample units* which in a geographical context are often areal units. The area population from which they are drawn is called the *target area*.

3.2 Timing of ground data collection in relation to remote sensing

Ground data can be collected before, during or after the collection of remote sensing data. Collection beforehand is usually pursued only for planning purposes, as in the location of suitable ground data sites: it is normally accompanied by the exploitation of archival data sources.

The collection of ground data synchronous with imaging is most desirable when the observed properties are dynamic (e.g. Bonn 1976). Thus crop surveys often require near simultaneous ground data collection, and studies of hydrological properties such as soil moisture, which display substantial diurnal variations, demand that the measurements are carefully timed (Jackson *et al.* 1976). Although such timing is ideal for interpretation of imagery, practical constraints often hinder its accomplishment. Manpower demands will be high and data collectors will probably need to be available over a period of time, since weather conditions will

39

often hinder imaging. Use of inexperienced personnel is quite feasible if only simple ground data are required (Townshend 1976), but for more sophisticated data collection relatively few skilled personnel are likely to be available. Automatic data recording is feasible for many ground attributes, but is usually expensive and is thus commonly restricted to research work (Curtis & Hooper 1974) in the context of remote sensing. A recent development has been the use of satellites for receiving data from ground stations and transmitting it to ground receiving stations. One type of data which should always be collected at the time of imaging, is that relating to meteorological conditions.

Frequently it is advisable to wait until the imagery has been collected successfully, and then to carry out the ground data collection as quickly as possible. Note should be made of recent changes since imaging, such as vegetation clearance. Thus ground data should not be collected during times of maximum rates of change whether these are phenological or a result of cropping (Joyce 1978).

For studies where time-invariant properties are most important it may be advisable to collect the ground data after imaging. This can be advantageous because preliminary analysis of the imagery improves sampling design, through scene stratification. In other words, ground sampling will usually be most efficient if the study area is subdivided into regions (i.e. areally stratified) and care then taken to sample each type of area. Such stratification can be most effectively executed by interpretation of the imagery accompanied by other data sources. Where this is done by human interpreters the resultant strata are likely to be of the land system or facet type (see Ch. 5). Unsupervised computer-assisted techniques (see Ch. 4.10) which create classes primarily on the basis of image properties are also preferably linked to data collection after sensing (Smedes *et al.* 1972, Nagy *et al.* 1971).

Ground data collection after sensing may involve the estimation of temporally variable ground properties. For example, satisfactory estimates of the ground cover of deciduous vegetation may involve judging the cover at times of year different from the date of imaging.

Timing of ground data collection will be influenced by external factors especially meteorological ones. High temperatures in arid areas during summer will lower work rates and areas without all-weather roads and with marked wet seasons as in the humid tropics, may be inaccessible for long periods. In arctic areas the effective field season may be even more limited.

3.3 Design of sampling schemes

Decisions about where to sample ground data are of great importance, since the validity of extrapolations about the whole of the target area by utilizing ground samples is highly dependent upon the design of the sampling scheme. Obtaining a sample from a larger population, whether carried out formally or informally, is an inherent component of all field work and its efficient execution is particularly important in reconnaissance surveys where the proportion of the area with ground data is likely to be very small.

Purposive sampling is where 'typical' sites are chosen subjectively. Proponents of such schemes argue that the resultant samples are very representative since they are based on the skills and possibly the local knowledge of the field worker. Furthermore, purposive sampling is effective in reducing the time expended on ground survey. Accessibility of sites in a target area will frequently be very variable, and thus selection of sample sites close to communication lines is likely to be

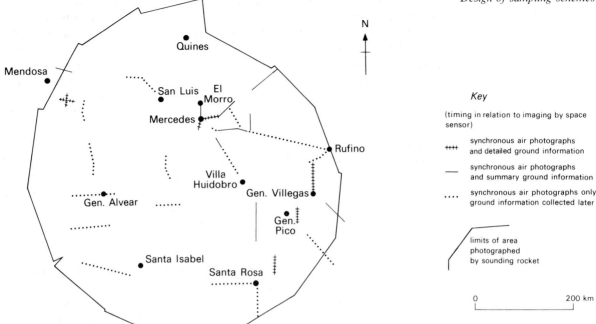

Key

(timing in relation to imaging by space sensor)

++++ synchronous air photographs and detailed ground information

——— synchronous air photographs and summary ground information

.... synchronous air photographs only ground information collected later

limits of area photographed by sounding rocket

0   200 km

Figure 3.1 Location of ground data traverses and aircraft traverses for a survey of Central Argentina using space imagery derived from a high altitude sounding rocket. An example of a multi-level purposive sample.

preferred though these may be untypical of areas more distant. Especially when low resolution space imagery is being used, precise ground location may be difficult away from roads, rivers or other notable landmarks, particularly where land cover types are separated by broad transition zones and where low angle slopes dominate the landscape. Linear traverses are one of the most common types of purposive sampling design, observations being made at such frequency and intensity as seems appropriate to the field worker. In particular, if aircraft images are collected simultaneously with higher-altitude ones, then linear traverses form the most practical system. Figure 3.1 shows the location of ground data traverses and aircraft traverses for a survey of central Argentina using a high-altitude Skylark rocket as a remote sensing platform for photographic cameras. Traverses were along roads and were chosen to cover as large a variety of terrain as possible (Townshend 1976).

The principal problem with purposive sampling is that a formal statement of the representativeness of the sample is unobtainable and thus the reliability of estimates for the whole of the target area is similarly unknown. These can only be properly assessed if **probability sampling** is used (Cochran 1963): statistical procedures used in automated classification (Ch. 4) are often based on the assumption that samples are randomly chosen. Probability sampling is 'a formal procedure for selecting one or more samples from the population in such a manner that each individual or sampling unit has a known chance of appearing in the sample' (Krumbein & Graybill 1965).

Normally the chances are made equal which leads to **simple random sampling**. In the simplest scheme, points are chosen using pairs of random numbers to select grid intersections and observations are then made at these points (Fig. 3.2a). If the emphasis is on spatial properties then a similar scheme is used to select sample areas (Bagwell *et al.* 1976) (Fig. 3.2b). An alternative to random sampling is *systematic sampling* where samples are taken from regularly spaced grid intersections (Fig. 3.2c). A random element can be introduced by aligning the grid randomly (Fig. 3.2d) (Bauer 1977). The advantage of systematic sampling is that the area will be

41

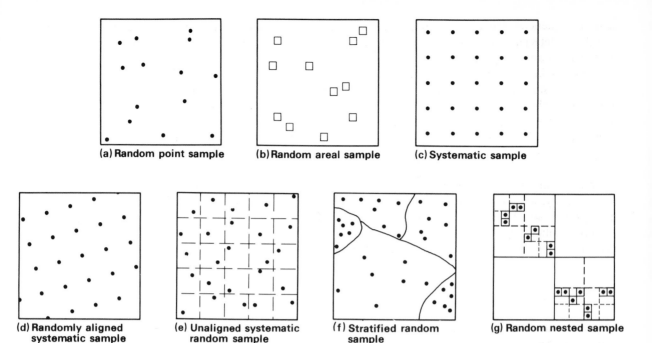

(a) Random point sample   (b) Random areal sample   (c) Systematic sample

(d) Randomly aligned systematic sample   (e) Unaligned systematic random sample   (f) Stratified random sample   (g) Random nested sample

Figure 3.2 Types of sampling schemes.

evenly covered: a specific disadvantage is that regularities in the terrain (as caused by aligned cuesta and strike vales) may coincide with the orientation of the systematic sample and lead to some areas being oversampled and others under-sampled. It is possible to combine the areal comprehensiveness of the systematic sample with the representativeness of the random sample by using an *unaligned systematic random sampling scheme*, where a random point or area is selected in every grid square (Fig. 3.2e). *Stratification* of the target area followed by sampling at different frequencies within each area is also commonly carried out (Fig. 3.2f), i.e. proportional sampling.

Despite the advantages of random sampling methods, they are used relatively rarely in ground sampling. This is a function of genuine practical problems of accessibility and accurate ground location which purposive sampling usually overcomes, but also because of a failure to recognize their benefits. In practice the authors' experience has been that random sampling schemes are less time consuming than is usually envisaged, so long as reasonably large scale air photographs are available and the terrain not too featureless. Even where full random sampling cannot be used, a random element can be introduced which will help reduce the distorting influence of individual type-site selection and help improve the representativeness of the sample. For example, random samples might be selected and the nearest similar site adjacent to an accessible road actually sampled. Alternatively sites could simply be chosen by random selection along roads or within a set distance from these access lines.

If sample units are large due to the use of low resolution satellite imagery (see next section), then a scheme of sampling within the unit needs to be devised to ensure that observations are representative of the unit as a whole. Such a scheme should be designed according to the property being estimated. Use of smaller areal subunits may be appropriate; alternatively line transects may be preferable if, for example, estimates of tree basal area are required. Goodspeed (1968) shows how, by making several measurements near a sample point and obtaining an average value, higher

42

spatial frequencies can be eliminated and the broader scale changes can more readily be detected. This has important implications in making sure that point observations are as representative as possible.

Reduction in travel times between randomly chosen sample sites can be achieved using a multistage *nested sampling scheme* (Lyon 1977). Areas are randomly chosen at one stage, then within these areas further ones are randomly selected and so on, until the final samples are selected (Fig. 3.2g). This scheme also potentially allows assessment of spatial variations in the relationships of ground characteristics with image properties (cf. Haggett 1965).

The design of a nested sampling scheme is often associated with multilevel sampling (Langley 1971, 1978), primary sampling units being first chosen on small-scale imagery, secondary sampling units on larger-scale imagery and so on until the final ground sampling units are selected (Rohde 1978). In a sense, the use of remote sensing is inherently multilevel, since at the very least it relies on data from a platform above the surface and on ground data.

3.4 Size and shape of sampling units

Most observations at ground level are made in areas rather than at points (cf. Alford *et al.* 1974) because of the nature of the characteristics to be measured. Therefore decisions need to be made about the areas' size and shape. The two principal factors controlling size are resolution of the imagery and inherent variability of the terrain.

The importance of resolution can be readily understood by reference to linescanner imagery, since the pixel provides a clear basic unit. Sample units smaller in area than a pixel will lead to unrepresentative observations of larger areas. Even sample units equal to the area of one pixel are not recommended, because of the problems of accurate ground location in terms of an external coordinate reference system, such as latitude and longitude, or a national grid (see final section in Ch. 2). Consider, that if ground location is determined to within plus or minus half a pixel, this still means that no single pixel-sized area on the ground can definitely be associated with one specific pixel. In general the minimum dimensions of the sample area, A, chosen may be estimated as follows:

$$A = P(1+2L)$$

where
    P = the pixel dimensions
    L = the accuracy of location in terms of number of pixels.

Thus with an accuracy of $\pm\frac{1}{2}$ pixel, the length and breadth of the sample unit using Landsat MSS data should be 158 m and for ± 1 pixel the size should be 237 m. This assumes that ground location can be determined accurately in the field. Where good topographic maps are absent, this may prove difficult and the minimum size of the sample area should be increased.

When other forms of imagery, such as photographs or radar, are used it is less easy to estimate a minimum size of unit which should be sampled, but it is necessary to avoid a choice of units too small to be detected and accurately located.

The difficulty of precise location also requires that sample units are internally homogeneous. Landscapes which consist of large areas of near homogeneous cover such as the large fields of the American mid-west provide abundant suitable sample units, though it is important that pixels falling across boundaries between fields or

Table 3.1 Criteria for considering a site to be uniform in an area of complex terrain (Justice 1978).

Characteristic	Allowable variability
Surface cover (I)	possesses one surface cover type for more than 85% of the site
Surface cover (II)	distribution of surface cover types must be spatially uniform throughout the site
Aspect	site must have no more than $22\frac{1}{2}$% variation either side of the dominant aspect
Slope angle	no more than 25% variation from dominant slope angle for no more than 20% of the site area

across ditches are rejected for training sets (Bauer 1977). Where a landscape is composed of many smaller units or where ground properties tend to change gradually rather than abruptly, as at field boundaries, it may be difficult to locate sample areas which are not noticeably heterogeneous. Decisions will then need to be made about the degree of internal variability allowable for an area to be regarded as homogeneous (Table 3.1). Additionally, where vegetation is comprised of mixtures of plant species these should have a reasonably even distribution throughout the sample (Fig. 3.3).

A more formal procedure for determining the minimum size of sample areas in regions of complex cover types has been suggested by Grabau and Rushing (1968), who define a 'structural cell' which is large enough fully to describe the vegetation's variability. If this size is greater than the size estimated in terms of multiples of pixel size then it should be used in preference. Clearly, followed to its logical conclusion, this will lead to the sample unit having variable size, depending on the type of vegetation (Beers 1978).

Similarly, for unvegetated surfaces the variability of soils in humid areas (Webster 1965, 1978, Campbell 1978) or of bare surfaces in arid areas needs to be considered (Mitchell *et al.* 1979). Since relief is often important in affecting remote sensing images, knowledge of its spatial variability (Stone & Dugundji 1965) is also relevant. Unfortunately, estimates of spatial frequencies of terrain characteristics are time consuming to make, and our knowledge of them is limited (Goodspeed 1968), though we know enough to be aware of the magnitude of their variability between different land types. Consequently, considerations of the spatial frequencies of land properties are difficult to include formally in a determination of sampling unit size.

The shape of sample areas is preferably rectangular or square when linescan imagery is used and for ease of location they are best orientated parallel to the scan lines. When using photographs a circular shape is equally preferable and is regarded by some as more natural (Grabau & Rushing 1968).

The number of samples which should be chosen is primarily dependent upon the variability of the land characteristics under assessment, the type of sensor–platform combination used, the effectiveness of the interpretation/classification procedure available and the required level of discrimination between different classes. Schemes for estimating sample sizes required in thematic mapping have been outlined by van Genderen *et al.* (1978) and Hay (1979) and in determining soil moisture by Rao (1976). Decisions about the required precision need to take account both of the costs of sampling and of the losses incurred by the user when the inventory information deviates from the true population values (Hamilton 1978), though there may well be

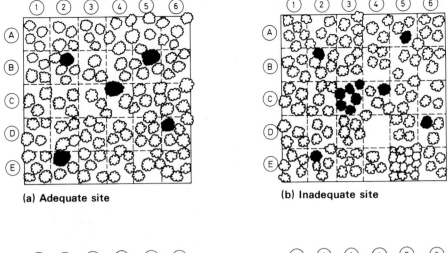

(a) **Adequate site**   (b) **Inadequate site**

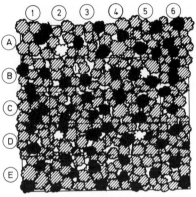

(a) **Adequate site**

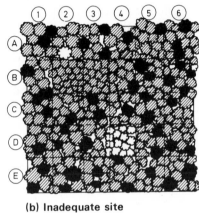

(b) **Inadequate site**

Figure 3.3 Selection of internally homogeneous sample units (from Joyce 1978).

substantial difficulties in estimating these costs and losses in monetary terms (Stuck & Burkhart 1978) and even greater problems when using non-economic criteria.

3.5 Selection of appropriate ground characteristics

3.5.1 *Types of properties*

We may distinguish three contrasting sets of properties on the basis of the objectives of ground survey.

(a) *Radiation characteristics at the ground surface.* These observations are important since they eliminate much of the interference due to the atmosphere and thus allow a readier identification of the relationships between ground characteristics and radiation properties. Such investigations may involve the use of sophisticated equipment (Fig. 3.4a), and have primarily been used for agricultural studies at test sites (e.g. LARS 1968, Miller *et al.* 1976). Commonly the equipment is non-imaging but instead provides accurate spectral data: the instrument shown in Figure 3.4a covers the range 0·62–2·5 μm (Silva 1978). Lee (1975) and Holmes (1970) provide

45

(a)

(b)

(c)

(d)

Figure 3.4 Types of ground radiation instruments. (a) Spectroradiometer of the Laboratory for Applications of Remote Sensing. Data are logged in the vehicle behind the 'cherry-picker' on which the radiometer is installed. (b) Band-pass radiometer operating in the field (designed by Edward Milton, Reading University). (c) Close-up of radiometer in (b). The sensor head is in the middle with its four sensor openings. Readings are taken from the meter on the left using the standard reference card so that reflectance values can be estimated. (d) Energy balance equipment for use with thermal infrared imaging. Sensible and latent heat fluxes can be obtained at five different heights.

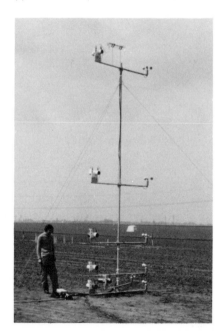

46

Table 3.2 Fundamental biophysical properties affecting radiation characteristics.

A *General*
   (a) Topography
   (b) Azimuth and elevation of radiation source
   (c) Intervening atmospheric interactions
   (d) Instrumental look angle
   (e) Internal instrumental characteristics

B *Visible and reflective infrared*

Vegetation	*Soils*
(a) Cellular structure (related to vigour, maturity and water stress)	(a) Soil texture
(b) Physical form of plant (e.g. leaf orientation, Leaf Area Index)	(b) Soil moisture
(c) Pigmentation	(c) Humus and iron oxide
(d) Foliar moisture content	(d) Mineralogical composition
	(e) Soluble salt content
	(f) Surface structure (microrelief)
	(g) Weathering products

C *Thermal infrared*

Basic physical controls:	*Thermal characteristics of surface dependent upon:*
(a) Temperature, dependent on energy balance (net short wave net long wave advection, evapotranspiration) and thermal properties such as thermal diffusivity, thermal inertia	(a) Soil properties (especially, moisture, texture, density and mineralogy) at surface and at depth
(b) Emissivity	(b) Vegetation properties
	(c) Near surface atmosphere properties (especially wind speed + humidity)

D *Microwave*
   (a) Microsurface roughness (and subsurface roughness)
   (b) Volume scattering (facets above surface in vegetation)
   (c) Complex dielectric constant (controlled particularly by moisture and conductivity)

useful reviews of such equipment. Of much interest in recent years have been easily portable band-pass radiometers which record the reflected radiation in the bands of the spectrum corresponding to particular channels, such as those of the multispectral scanners on board the Landsat series (Tucker 1978, Milton 1979) (Figs 3.4b, c). Such instruments, although providing fewer data than the continuous spectral readings of more sophisticated instruments, can undoubtedly greatly aid interpretation of sensor data; they are relatively inexpensive and they can therefore be incorporated readily into operational surveys. Similar field instruments are obtainable for recording thermal infrared radiation, such as the Barnes P R T-5 and P R T-10 (Bonn 1976); field estimates of the components of the energy balance require more sophisticated equipment (Fig. 3.4d). Field instruments for use in the microwave region are described by Chapman *et al.* (1970) and Bryan and Larson (1973).

(b) *Fundamental biophysical properties affecting radiation characteristics.* An understanding of the factors controlling the radiation leaving the Earth's surface often demands a consideration of fundamental biophysical properties and hence rigorous measurement and instrumentation can be necessary. The properties 47 selected for measurement will be strongly affected by the type of radiation. An illustration of this is the roughness of ground surface which affects whether radiation is specular or diffuse. Rough surfaces which cause diffuse radiation can be defined using the Rayleigh criterion (Chapman *et al.* 1970):

$$h < \frac{\lambda}{8 \sin \gamma}$$

where
h = height of surface irregularities
λ = wavelength of the electromagnetic radiation
γ = incident angle.

Thus the sizes of features which are relevant for visible radiation are very much smaller than for microwave radiation, and within the latter the size will alter substantially according to wavelength.

Table 3.2 describes some of the main physical properties affecting radiation recorded for different parts of the spectrum. Although clearly not a part of most operational land resources surveys, studies of the fundamental controls affecting radiation form the basis of our understanding of the capability of remote sensing and an appreciation of their significance is necessary for a continued improvement of operational systems.

(c) *Directly observable field properties.* In operational surveys the ground data collected are primarily for extrapolations over the whole target area and may include some of the properties in Table 3.2. The properties are usually those needed to describe the final classes or properties in the terrain analysis. The objectives of the survey will determine the choice of the properties that are selected and their variety makes it impossible even to outline the range of possibilities. Some sets of characteristics have a direct effect on the appearance of remotely sensed data, namely those of surface form, surface materials and land cover. These are important not only in their own right but also since they form the basis for inferences about other sets such as geological, geomorphological, hydrological and pedological ones (Section 4.7) The three basic sets of characteristics are outlined in turn below.

3.5.2 *Surface form characteristics*

Slope angle and aspect are the principal terrestrial determinants causing areas either to be directly illuminated by the Sun or to be in shadow and hence to receive only diffuse sky radiation. The interaction of Sun angle and azimuth with slope angle and aspect in causing variations in surface illumination is given by an incidence value (Monteith 1973, p. 41):

incidence value $= \cos\alpha + \sin\alpha.\cot\beta.\cos\theta$
where
β = solar elevation from the horizontal
α = slope angle
θ = difference between slope aspect and solar azimuth.

The recording of these properties is often essential for interpretation of other surface properties because of their effect on the radiance received at the sensor (Ellis 1978), which is a function of the incidence and exitance angles (Holben & Justice

1979). Hoffer and staff (1975) also included mean elevation in an 'effective illumination index'. Other purely morphological properties are the profile form, namely whether convex, concave or straight and combinations of these forms (Savigear 1956), as well as the plan form, represented by symbolic representation of breaks of slope (Savigear 1965) to produce a morphological map. The location of sites for detailed sampling may usefully be represented by means of a morphological map (Curtis *et al.* 1965). At a much larger scale is the recording of microrelief, which apart from its relevance as a terrain descriptor, has a direct effect on reflected radiation especially in the microwave part of the spectrum (Lee 1975).

Purely morphological descriptions should be distinguished from geomorphological ones, the latter including the origin of the forms. Similarly, Savigear (1965) distinguishes between 'land forms' and 'landforms' as the subject of the two types of description. Several comprehensive schemes have been devised for geomorphological mapping (e.g. Tricart 1965, Verstappen & van Zuidam 1968, Demek 1972, van Dorsser & Salome 1973, 1974, Demek & Embleton 1978), and these can be exploited to develop an individual key for a particular survey. Although the recording and mapping of geomorphological features is a valuable aid in many environmental management problems (Cooke & Doornkamp 1974), it is often a complex technical task, and the recorded information is not necessarily easily extrapolated to the whole of a target area from sample units.

3.5.3 *Surface material variables*

A full description of surface materials should include estimates of texture with field descriptions of the mineralogical composition of particles coarser than silts (Soil Survey Staff 1951, Clarke 1971). Organic matter should be recorded such as the amount of litter and types of humus, though laboratory testing will be necessary for reliable estimates of the percentage content. Soil colour can be recorded using a Munsell colour book but a band-pass radiometer may be preferable (Section 3.5.1). Soil moisture can be estimated by eye and touch (Soil Survey Staff 1951) but accurate measurements require either the removal of materials for gravimetric calculation or the use of more sophisticated ground instrumentation for accurate monitoring, such as neutron probes. The presence of secondary accumulations such as carbonates, and salts like gypsum needs to be noted. The degree of compaction and the soil structure as represented by the size and shape of aggregates at the surface should be described, these properties overlapping at their largest with microrelief at its smallest. A comprehensive characterization of soils requires details of the component horizons (Hodgson 1978), either through augering or less frequently by digging pits. It may be necessary then to classify the soil according to one of the principal available classifications (e.g. Soil Survey Staff 1975, FAO-Unesco 1974, Avery 1973) though this can be time consuming and should not be undertaken lightly. In more arid areas no true soils may be present, but only infrequently will bare rock be exposed at the surface, and the description of the surface materials will include several of the properties used to describe soils proper. Where bare rock is exposed, its lithology should be recorded (Lahee 1960), with data on the degree and type of surface weathering.

Many of these properties will be relevant to several application areas but differences in terminology, their relative significance and the choice of the relevant class intervals will vary according to the objectives of a survey. Several of the sources quoted above have been written with agriculture specifically in mind (e.g. Soil Survey Staff 1951): properties relevant to engineering are outlined by the

International Association of Engineering Geology (1976) and USDI (1974). For much engineering work sophisticated field tests are necessary for accurate estimates of characteristics (e.g. Kovari 1977), but there are schemes for obtaining simple field descriptions of properties such as plasticity, dry strength and dilatancy (FAO 1973).

As well as their inherent importance, surface materials are significant because of their effect on the appearance of vegetation when the cover is incomplete (Richardson 1975, Gerbermann *et al.* 1976). Estimation of the percentage bare ground often needs to be made (Driscoll 1972; Musick 1973). Cover of low herbaceous vegetation may be made accurately by quadrat, though this is time consuming and after some training direct visual estimates can often be made with sufficient accuracy. Ground cover of tree vegetation determined from ground level has to be estimated visually: but low-level air photographs can give reliable estimates (Gerbermann *et al.* 1976) as can vertical stereo-pairs of photographs of herbaceous vegetation taken from ground level or a raised platform.

3.5.4 *Vegetation cover properties*

Vegetation description can be lengthy and complex, so only those properties that are really needed should be considered. Fosberg (1967) distinguished between physiognomic, structural, functional and compositional criteria for vegetation classification. Using his definitions (Table 3.3), although a physiognomic description can be useful, a structural one is likely to be more valuable, since the physical properties of the vegetation are described more precisely. Height classes distinguishing between tree shrubs and a herbaceous layer are commonly made (Table 3.4), though the presence of woody tissue may also be included in the distinction between

Table 3.3 Types of criteria for vegetation classification (Fosberg 1967).

Types of criteria	*Characteristics*
Physiognomic	refers to appearance, particularly external. Separates large categories like forest, grassland, and savanna. Includes features like seasonality and luxuriance.
Structural	arrangement in space of vegetation. Includes height, branching habit, size of stems and crowns, density of canopy, stratification.
Functional	features adapted to environment. Includes xeromorphy, means of resistance to fire, halomorphy, dispersal mechanisms.
Compositional	species of vegetation; usually includes proportional composition.

Table 3.4 Structural vegetation classes.

Class	*Height*
Herbaceous	less than 1·3 m
Shrub	1·3 – 3 m
Tree	greater than 3 m

Class	*Spacing*
Closed	predominantly touching
Open	predominantly discrete
Sparse	more than twice average diameter apart

shrub and herb (Fosberg 1967). Where more than one layer is significant their relative importances must be indicated. These can be expressed in terms of the percentage cover of each layer projected down to the ground surface, though because of overlapping layers, totals greater than 100% can arise. The leaf area index (LAI) is the cumulative one-sided leaf area per unit ground area from the top surface of a stand to a plane usually at ground level (Janza *et al.* 1975). Alternatively, one may more simply measure the percentage area projected on to a plane above the vegetation. The spatial arrangement of the vegetation as a whole (Fosberg 1967) or of individual layers can be described in terms of crown spacing. Functional criteria are less relevant for our purposes since they describe adaptations to the environment which in terms of the overall appearance of plants may be relatively small, such as the height of budding: but they may be significant since they include times of flowering and fruiting. Description of the composition in terms of species is an ideal which frequently will not be achieved because of the high demands on time and expertise, though for more specialized geobotanical investigations it may be essential (Cole *et al.* 1974). It may prove useful to assign an area of vegetation to a particular vegetation type (Dansereau 1958, Kuchler 1949, Fosberg 1967) such as an association or formation. Christian and Perry (1953) suggest a symbolic system for describing vegetation: features such as diseases also need to be noted if prevalent, along with maturity though the criteria for the latter will vary greatly depending on the sort of vegetation, viz. whether a herb or tree (Joyce 1978). Surveys for particular purposes will demand the collection of specialized data as, for example, is described by the Soil Conservation Service of the US Department of Agriculture for native grazing lands (Soil Conservation Service 1976). Surveys including crops will require recording of other specific characteristics (LARS 1967, 1968; Curtis & Hooper 1974; Townshend 1976) such as dates of planting, planting techniques and row orientation. Land use as opposed to land cover (Section 1.2.3) should also be recorded separately: Poulton (1972) gives a useful key for the ground description of both natural and altered landscapes.

Some types of terrain are sufficiently distinctive to demand a separate scheme of description: Adams and Zoltai (1969) suggest a comprehensive one for open water and wetlands and obviously coastlines will also warrant separate consideration (King 1974).

3.6 Systematic recording of ground data

The collection of ground data is facilitated by the production of a standard ground data form, with the provision of either sections to be filled in or with a list of items that can be checked off. In this way standardized data will be assembled for every sample unit and generalizations on the basis of the units more readily realized. Although the design of ground data forms is a rather prosaic activity, it is nonetheless important and the effectiveness of such forms should be tested thoroughly before being used operationally. This is to ensure that all necessary data are collected and that they can be collected easily and with a low error rate. Some preliminary training of ground data collectors is highly desirable. Form design is dependent upon the complexity and type of data to be collected and within the same region different conditions may demand the use of different forms. Both Heyligers (1968) and Joyce (1978) propose different forms depending upon the type of vegetation community, for example whether forest or shrub. Usually the forms should be compact for ease of handling in the field but conversely there should be adequate space for necessary

51

```
┌──────────────────────────────────────────────────────────────────┐
│ GENERAL DATA                                                       │
├──────────────────────────────────────────────────────────────────┤
│ DATE:                                SITE NUMBER:                  │
├──────────────────────────────────────────────────────────────────┤
│ Terrain unit:                        Grid reference:              │
│ Map sheet:                                                         │
├──────────────────────────────────────────────────────────────────┤
│ Aerial photo number:                 Description of location in   │
│                                      relation to other units:     │
│ Ground photo number:    overview:                                 │
│                         internal:                                 │
└──────────────────────────────────────────────────────────────────┘
```

```
┌──────────────────────────────────────────────────────────────────┐
│ SITE DESCRIPTION                                                   │
├──────────────────────────────────┬─────────────────────────┬──────┤
│ Aspect:                          │ Hydrological conditions │ S R F N│
├──────────────────────────────────┼─────────────────────────┤      │
│ Slope angle:                     │ Surface material class: │      │
│ Ht. a.s.l. (m):                  │ Soil texture:           │      │
│ Microrelief:                     │ Depth (cms):            │      │
│ % unvegetated surface:           │ Parent material:        │      │
│                                  │                         │      │
│ Landform description:            │ Soil colour:            │      │
│                                  │ Organic matter:         │      │
│                                  │ Surface structure:      │      │
└──────────────────────────────────┴─────────────────────────┴──────┘
```

VEGETATION DESCRIPTION
% vegetation cover: Vegetation class:
% evergreen vegetation: Secondary cover:
% deciduous vegetation: Spatial distribution:
% tree cover: % shrub cover: % herbaceous cover:

Species data

	Name	% cover	% cover (predicted for different seasons)	Ht. (m)
1				
2				
3				
4				
5				
6				
7				
8				
9				
10				

Arable rotation information: Surface cover class:
Land use:
(Additionally a cross-section showing vegetation structure should be drawn and a
  block diagram showing three-dimensional form in relation to surrounding sites)

Figures 3.5 Example of a ground data form for land cover description.

annotation bearing in mind difficulties of writing in the field. An example of a field information form for a general purpose land cover survey is given in Figure 3.5.

There is a temptation in designing forms to include unnecessary space for rare phenomena on the off-chance they will be important in interpretation. In practice forms should be pruned rigorously to include only those variables that are really essential. Conversely it can be a mistake to categorize data into very broad classes. If relatively high precision can be achieved easily there is little to be gained by assigning these values to broad classes, a procedure readily done in subsequent computerized analysis. If data at higher accuracy are ever required in the future they are then still extractable. Inclusion of appropriate spaces for subsequent coding to facilitate storage on computer cards or tapes is strongly recommended.

In a large data collection programme it is inevitable that errors will be made in filling out forms. Thus it is useful to take ground level photographs of each site visited, to act as a check of ground data forms if any possible discrepancies are discovered. Where ground data collection occurs after the images are obtained, and where the quantity of data collected for a given site is small, then direct annotation of the images with symbolic notation can prove very efficient.

3.7 Conclusions

We have outlined various stages so that ground data are sampled at the right time, at appropriate localities, in areal units appropriate in size and shape for the images used and for the terrain being investigated: also that the appropriate properties are chosen and that they are recorded in a form suitable for subsequent analysis. There remains the task of relating the ground data to the remotely sensed data, a task which strongly affects the accuracy of the whole terrain analysis.

If the size of the areal units is large in relation to the scale of the images used, then practice readily demonstrates that little difficulty is found in establishing a correspondence between the two data sets. But where this is not so, as for small-scale space imagery, significant locational difficulties can arise. Lower-level, higher-resolution images can be of substantial assistance in this respect. For example, the patterns perceived on air photographs can often easily be related to those recognized on enlarged Landsat images even if they were taken much earlier. Ground sample units located on the air photographs can thus be related to the correct part of the image. This is greatly facilitated by an optical instrument such as a Bausch Lomb Transferscope (Downs *et al.* 1977) or OMI Stereo-facet plotter (Fig. 3.6) where two images can be overlain optically using controls which alter their relative scales and their geometries.

If ground location can be accurately determined on a topographic map, then least squares transformation of digital imagery can be used to locate ground sample points (Section 2.6). Eppler and Merrill (1968) describe an automated method by which aircraft linescanner imagery can be related directly to topographic maps.

The paucity of work on ground data collection schemes has hindered the expansion in use of remote sensing methods, by leading both to over- and to underestimates of their value. Extravagant claims for remote sensing have often been unsupported by ground verification. Conversely, the lack of emphasis on thorough investigations of the effects of ground properties on remote sensing data may well have led to potentially useful relationships being underexploited.

An implicit assumption of this chapter is that ground data collection should be carried out systematically and carefully at all stages. Without such thoroughness, the validity of all aspects of terrain resources survey must be thrown into doubt. But, however much care is taken, the resultant observations will inevitably fail to be totally representative of the population. Smedes (1975) notes the virtual impossibility of producing a truly accurate depiction of ground conditions because of the spatial

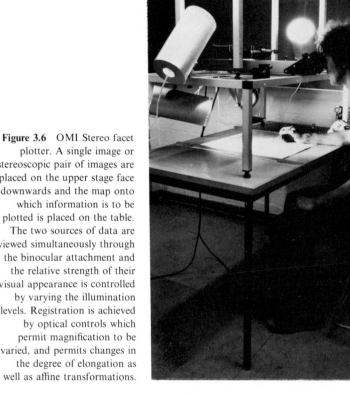

Figure 3.6 OMI Stereo facet plotter. A single image or stereoscopic pair of images are placed on the upper stage face downwards and the map onto which information is to be plotted is placed on the table. The two sources of data are viewed simultaneously through the binocular attachment and the relative strength of their visual appearance is controlled by varying the illumination levels. Registration is achieved by optical controls which permit magnification to be varied, and permits changes in the degree of elongation as well as affine transformations.

complexity of terrain, the problem of mixed terrain classes and the problem of boundaries. He points out that, especially in terms of spatial extent, the imagery may in a sense be more accurate than the ground data. Clearly this implies a reliance on the integration of ground level and remotely sensed data, using the strengths of both. An expansion of research work on all aspects of ground data collection is necessary for a more productive use of remote sensing imagery in land resources survey.

References

Adams, G. D. and S. A. Zoltai 1969. Proposed open water and wetland classification. In D. S. Lacate, 1969. *Guidelines for biophysical land classification*, 23–41. Ottawa: Queens Printer.

Alford, M., P. Tuley, E. Hailstone and J. Hailstone 1974. The measurement and mapping land resource data by point sampling on aerial photographs. In *Environmental remote sensing*, E. C. Barrett and L. F. Curtis (eds), 113–26. London: Edward Arnold.

Avery, B. W. 1973. Soil classification in the soil survey of England and Wales. *J. Soil Sci.* **24**, 324–38.

Bagwell, C., G. C. Sharma and S. W. Downs 1976. Ground truth study of a computer generated land use map of North Alabama. In *Remote sensing of Earth resources*, **5**, F. Shahrokhi (ed.), 139–50. Tullahoma, Tennessee: University of Tennessee.

Bauer, M. 1977. *Crop identification and area estimation over large geographic areas using Landsat MSS data.* LARS Tech. Rep. 012477.

Beers, T. W. 1978. Developing efficient estimation techniques for integrated inventories. *Proc. Workshop on Integrated Inventories of Renewable Natural Resources, Tucson, Arizona*, 270–5.

Belcher, D. J. 1948. Determinations of soil conditions from aerial photographs. *Photogramm. Engng* **14**, 482–8.

Benson, A. S., W. C. Draeger and L. R. Pettinger 1971. Ground data collection and use. *Photogramm. Engng* **37**, 1159–67.

Bonn, F. 1976. Some problems and solutions related to ground truth measurements for thermal infra-red remote sensing. *Proc. Am. Soc. Photogramm.* **42**, 1–11.

Bryan, M. L. and R. W. Larson 1973. Application of dielectric constant measurements to radar imagery interpretation. In *Remote sensing of Earth resources*, **2**, F. Shahrokhi (ed.), 529–48. Tullahoma, Tennessee: University of Tennessee.

Buringh, P. 1960. The application of aerial photographs in soil surveys. *Manual of photographic interpretation*, 633–66. Washington, DC: American Society of Photogrammetry.

Campbell, J. B. 1978. Spatial variation of sand content and pH content within single contiguous delineations of two soil mapping units. *Soil Sci. Am. J.* **42**, 460–4.

Chapman, P., J. Quade, P. Brennan and J. C. Blinn 1970. Ground truth/sensor correlation. *2nd Annual Earth Resources Program Rev., NASA, Houston, Texas*, **2**, Section 31.

Christian, C. S. and R. A. Perry 1953. The systematic description of plant communities by the use of symbols. *J. Ecol.* **41**, 100–105.

Clarke, G. R. 1971. *The study of soil in the field.* Oxford: Oxford University Press.

Cochran, W. G. 1963. *Sampling techniques*, 2nd edn. New York: John Wiley.

Cole, M. M., E. S. Owen-Jones, N. D. E. Custance and T. E. Beaumont 1974. Recognition and interpretation of spectral signatures of vegetation and satellite imagery in Western Queensland, Australia. *Proc. Symp. on Eur. Earth Resources Satellite Experiments*, ESRO SP-100, 243–87.

Cooke, R. U. and J. C. Doornkamp 1974. *Geomorphology in environmental management.* Oxford: Oxford University Press.

Curtis, L. F., J. C. Doornkamp and K. J. Gregory 1965. The description of relief in field studies of soils. *J. Soil Sci.* **16**, 16–30.

Curtis, L. F. and A. J. Hooper 1974. Ground-truth measurements in relation to aircraft and satellite studies of agricultural land use and land classification in Britain. *Proc. Symp. on Eur. Earth Resources Satellite Experiments*, ESRO SP-100, 405–15.

Dansereau, P. 1958. *A universal system for recording vegetation.* Canada: Institut Botanique de l'Université de Montreal.

Demek, J. (ed.) 1972. *Manual of detailed geomorphological mapping.* Prague: Academia.

Demek, J. and C. Embleton 1978. *Guide to medium-scale geomorphological mapping.* Stuttgart: E. Schweizerbart'sche Verlagsbuchhandlung.

Downs, S. W., G. C. Sharma and C. Bagwell 1977. *A procedure for a ground truth study of a land use map of North Alabama generated from Landsat data.* NASA Tech. Note, NASA TN D-8420.

Driscoll, R. S. 1972. Multi-spectral scanner imagery for plant community classification. *Proc. 8th Int. Symp. on Remote Sensing of Environment, Ann Arbor, Michigan*, 1259–78.

Ellis, S. L. 1978. Shrubland classification in the Central Rocky Mountains and Colorado Plateau. *Proc. Natl Workshop on Integrated Inventories of Renewable Natural Resources, Tucson, Arizona*, 199–203.

Eppler, W. G. and R. D. Merrill 1968. *Correlating remote sensor signals with ground-truth information.* Alto, California: Lockheed Missiles and Space Co.

FAO 1973. Soil survey interpretation for engineering purposes. *Soils Bull.* no. 19.

FAO-Unesco 1974. *Soil map of the world*, vol. 1 legend. Rome: FAO.

Fosberg, F. R. 1967. A classification of vegetation for general purposes. In *Guide to the checksheet for IBP handbook* **4**, G. F. Peterken (ed.), 73–120. Oxford: Blackwell.

55

Gerbermann, A. M., J. A. Cuellas and C. L. Wiegand 1976. Ground cover estimated from aerial photographs. *Photogramm. Engng* **42**, 551–6.

Goodspeed, J. 1968. Sampling considerations in land evaluation. In *Land evaluation*, G. A. Stewart (ed.), 40–52. Melbourne: Macmillan.

Grabau, W. E. and W. N. Rushing 1968. A computer-compatible system for quantitatively describing the physiognomy of vegetation assemblages. In *Land evaluation*, G. A. Stewart (ed.), 263–75. Melbourne: Macmillan.

Haggett, P. 1965. *Locational analysis in human geography*. London: Edward Arnold.

Hamilton, D. A. 1978. Specifying precision in natural resource inventories. *Proc. Workshop on Integrated Inventories of Renewable Natural Resources, Tucson, Arizona*, 276–81.

Hay, A. M. 1979. Sampling designs to test land-use map accuracy. *Photogramm. Engng and Remote Sensing* **45**, 529–34.

Heyligers, P. C. 1968. Quantification of vegetation structure on vertical aerial photographs. In *Land evaluation*, G. A. Stewart, (ed.), 251–62. Melbourne: Macmillan.

Hittle, J. E. 1949. Airphoto-interpretation of engineering sites and materials. *Photogramm. Engng* **15**, 589–603.

Hodgson, J. M. 1978. *Soil sampling and soil description*. Oxford: Oxford University Press.

Hoffer, R. M. and staff 1975. *Computer-analysis of Skylab multi-scanner data in mountainous terrain for land use, forestry, water resource and geological applications*. LARS Inf. Note 121275.

Holben, B. N. and C. O. Justice 1979. *Evaluation and modelling of the topographic effect on the spectral response from nadir-pointing sensors*. NASA/GSFC Tech Mem. 80305.

Holmes, R. A. 1970. Field spectroscopy. In *Committee on Remote Sensing for Agricultural Purposes: remote sensing with special reference to agriculture and forestry*, 298–323. Washington, DC: National Academy of Sciences.

International Association of Engineering Geology. 1976. *Engineering geology maps*. Paris: Unesco.

Jackson, R. D., R. S. Reginato and S. B. Idso 1976. Timing of ground truth acquisition during remote sensing assessment of soil water content. *Remote Sensing of Environment* **4**, 249–56.

Janza, F. J. *et al.* 1975. Interaction mechanisms. In *Manual of remote sensing*, R. G. Reeves (ed.), 75–179. Virginia: American Society of Photogrammetry.

Joyce, A. T. 1978. *Procedures for gathering ground truth information for a supervised approach to a computer-implemented land cover classification of Landsat-acquired multispectral scanner data*. NASA Ref. Publ. 1015.

Justice, C. O. 1978. *The effect of ground conditions on Landsat multispectral scanner data for an area of complex terrain in southern Italy*. Unpub. PhD Thesis, Reading University, UK.

King, C. A. M. 1974. Coasts. In *Geomorphology in environmental management*, R. U. Cooke and J. C. Doornkamp (eds), 188–222. Oxford: Oxford University Press.

Kovari, K. 1977. *Field measurements in rock mechanics*. Rotterdam: Balkema.

Kuchler, A. W. 1949. A physiognomic classification of vegetation. *Ann. Assn Am. Geogrs* **39**, 201–10.

Lahee, F. H. 1960. *Field geology*. New York: McGraw-Hill.

Langley, P. G. 1971. Multistage sampling of earth resources with aerial and space photography. In *Monitoring Earth resources from aircraft and spacecraft*, 129–41. NASA SP-275. Washington DC: NASA.

Langley, P. G. 1978. Remote sensing in multi-stage, multi-resource inventories. *Proc. Natl Workshop on Integrated Inventories of Renewable Natural Resources.' Tucson, Arizona*, 205–208.

LARS (Laboratory for Agricultural Remote Sensing) 1967. *Remote sensing in agriculture*. 2nd Annual Rep. of LARS, West Lafayette, Indiana.

LARS 1968. *Remote sensing in agriculture*. 3rd Annual Rep. of LARS, West Lafayette, Indiana.

Lee, K. 1975. Ground investigations in support of remote sensing. In *Manual of remote sensing*, R. G. Reeves (ed.), 805–56. Virginia: American Society of Photogrammetry.

Lintz, J., P. A. Brennan and P. E. Chapman 1976. Ground-truth and mission operations. In

Remote sensing of environment, J. Lintz and D. S. Simonett (eds), 412–36. Reading, Mass.: Addison-Wesley.

Lyon, J. P. 1977. *Evaluation of ERTS MSS signatures in relation to ground control signatures using a nested sampling approach.* NASA ERTS Rep. E 77-10051, final rep.

Miller, L. D., R. L. Pearson and C. J. Tucker 1976. A mobile field spectrometer laboratory. *Photogramm. Engng and Remote Sensing* **42**, 569–72.

Milton, E. 1979. *Reading band-pass radiometer user's manual.* Remote Sensing Rep., Reading University, UK.

Mitchell, C. W., R. Webster, P. H. T. Beckett and B. Clifford 1979. An analysis of long-range predictions of conditions in deserts. *Geogl J.* **145**, 72–85.

Monteith, J. L. 1973. *Principles of environmental physics.* London: Edward Arnold.

Musick, H. B. 1973. *A study to explore the use of orbital remote sensing to determine native land plant distribution.* NASA ERTS Rep. E73–10367.

Nagy, G., G. Shelton and J. Tolaba 1971. Procedural questions in signature analysis. *Proc. 7th Int. Symp. on Remote Sensing of Environment, Ann Arbor, Michigan,* 1387–401.

Pettinger, L. R. 1971. Field data collection – an essential element in remote sensing applications. *Proc. Int. Workshop on Earth Resources Survey Systems, Washington, DC,* 49–64.

Poulton, C. E. 1972. A comprehensive remote sensing legend system for the ecological characterization and annotation of natural and altered landscapes. *Proc. 8th Int. Symp. on Remote Sensing of Environment, Ann Arbor, Michigan,* 393–408.

Rao, R. G. S. 1976. *Sampling techniques for ground truth data acquisition in microwave remote sensing of soil moisture.* Remote Sensing Lab. Tech. Rep. 264–11, University of Kansas, Lawrence.

Richardson, A. J. 1975. Plant, soil and shadow reflectance components of row crops. *Photogramm. Engng and Remote Sensing* **41**, 1401–407.

Rohde, W. G. 1978. Potential applications of satellite imagery in some types of natural resource inventories. *Proc. Natl Workshop on Integrated Inventories of Renewable Natural Resources, Tucson, Arizona,* 209–18.

Savigear, R. A. G. 1956. Technique and terminology in the development of the investigation of slope forms. *Slopes Comm. Rep.* **1**, 66–75.

Savigear, R. A. G. 1965. A technique of morphological mapping. *Ann. Assn Am Geogrs* **55**, 514–38.

Silva, L. F. 1978. Radiation and instrumentation, an overview. In *Remote sensing: the quantitative approach,* P. H. Swain and S. M. Davis (eds), 21–135. New York: McGraw-Hill.

Smedes, H. W. 1975. The truth about ground truth. *Proc. 10th Int. Symp. on Remote Sensing of Environment, Ann Arbor, Michigan,* 821–3.

Smedes, H. W., H. L. Linnerud, L. B. Woodlaver, M. Y. Su and R. R. Jayroe 1972. Mapping terrain data by computer clustering techniques using multispectral scanner data and using color aerial films. *4th NASA Earth Resources Program Rev.* **3**, 61-1 to 61-30.

Soil Conservation Service 1976. *National range handbook.* Washington, DC: USDA.

Soil Survey Staff 1951. *Soil survey manual.* USDA Handbook 18. Washington, DC: USDA.

Soil Survey Staff 1975. *Soil taxonomy.* Soil Conservation Service. Agricultural Handbook 436. Washington, DC: USDA.

Stone, R. O. and J. Dugundji 1965. A study of microrelief – its mapping, classification and quantification by means of a Fourier analysis. *Engng Geol.* **1**, 1–89.

Stuck, D. R. and H. E. Burkhart 1978. A framework for allocating inventory resources for multiple-use planning. *Proc. Natl Workshop on Integrated Inventories of Renewable Natural Resources, Tucson, Arizona,* 307–14.

Swain, P. H. and S. M. Davis (eds) 1978. *Remote sensing: the quantitative approach.* New York: McGraw-Hill.

Townshend, J. R. G. 1976. Ground information for the earth resources Skylark. In *Environmental remote sensing* **2**, E. C. Barrett and L. F. Curtis (eds), 217–45. London: Edward Arnold.

Tricart, J. 1965. *Principes et méthodes de la géomorphologie.* Paris: Masson.

57

Tucker, C. J. 1978. *An evaluation of the first four Landsat-D thematic mapper reflective sensors for monitoring vegetation: a comparison with other satellite sensor systems.* NASA Tech. Mem. 79617. Greenbelt, Md: Goddard Space Flight Center.

Tucker, C. J. and E. L. Maxwell 1976. Sensor design for monitoring plant canopies. *Photogramm. Eng. and Remote Sensing* **42**, 1399–410.

USDI (United States Department of the Interior) 1974. *Earth manual*, 2nd edn. Washington, DC: USDI

van Dorsser, H. J. and A. I. Salome 1973. Different methods of detailed geomorphological mapping. *KNAG Geografisch Tijdschift* **7**, 71–4.

van Dorsser, H. J. and A. I. Salome 1974. Two methods of detailed geomorphological mapping. *KNAG Geografisch Tijdschift* **8**, 467–8.

van Genderen, J. L., B. F. Lock and P. A. Vass 1978. Remote sensing: statistical testing of thematic map accuracy. *Remote Sensing of Environment* **7**, 15–36.

Verstappen, H. Th. and R. A. van Zuidam 1968. ITC system of geomorphological survey. *ITC Textbook of photo-interpretation*, Ch. VII, 2. Delft: ITC.

Vink, A. P. A. 1968. Aerial photographs and the soil sciences. *Proc. Conf. on Aerial Surveys and Integrated Studies, Toulouse*, 81–141. Paris: Unesco.

Webster, R. 1965. *Minor statistical studies on terrain evaluation.* MEXE Rep. 877, Christchurch, England.

Webster, R. 1978. Mathematical treatment of soil information. *Proc. 11th Conf. Int. Soc. Soil Sci., Edmonton, Canada*, **3**, 161–90.

4 Image analysis and interpretation for land resources survey

John R. G. Townshend

4.1 Introduction

4.1.1 *Human interpretation and machine-assisted methods*

Newer methods of data handling, processing and classification offer substantial opportunities to those concerned with terrain resources evaluation. Such methods have been adopted far less readily than the newer methods of data collection by remote sensing (Verstappen 1974), as described in Chapter 2. Analyses of these data have most frequently been carried out by human interpreters adapting the well-tried methods of photo-interpretation. Such methods have proved successful but it is increasingly apparent that a variety of techniques are available which can assist, and for some functions even largely replace, the human interpreter.

Justification for at least a partial replacement of the human interpreter is twofold. Firstly, the quantity of data in image form has increased substantially, principally through the development of new remote sensing data collection systems such as Landsat. Furthermore, the dimensionality of the data has increased: the multispectral scanners in Landsats 1 and 2 had four separate bands, that in Landsat 3 had five bands and the Thematic Mapper of Landsat D is a multispectral scanner with seven bands. The human interpreter, therefore, needs assistance in coping with this increased data flow. Secondly, as will be shown below, abstraction of the full value of these data can often only be obtained if they are subjected to image processing. Improvements in data handling can lead both to faster and more accurate interpretations and to interpretation at larger scales.

Nevertheless, new techniques bring with them new problems and restrictions: the methods to be discussed bring no panacea, and discussion of the limitations of the various techniques will be included as they are described. In this chapter emphasis on the newer methods of image interpretation reflects our interest in their potential and not a failure to appreciate the merits of human interpretation skills which are described more fully elsewhere (e.g. Lueder 1959, American Society of Photogrammetry 1960, Miller & Miller 1961, von Bandat 1962).

4.1.2 *Image enhancement and classification*

It is convenient to distinguish between *image enhancement* and *classification*, which together form the two main components of *image processing*. The former is composed of transformations and manipulations in order to facilitate analysis and classification. Frequently the term *preprocessing* is used in the literature, though without noticeable consistency. We suggest it can usefully be restricted to describe the transformation of images to create a more accurate depiction of the scene in terms of its radiation characteristics and geometric properties (Section 2.6). The former includes many aspects of image restoration, as well as coding and bandwidth compression (Hunt 1975), which are also important in image transmission and storage. Correction for atmospheric effects (e.g. Fraser 1974, Parsons & Jurica 1975) by this definition also forms part of image preprocessing. In practice the dividing line between preprocessing and processing can be very diffuse. *Classification* is the set of actions which leads to parts of the image being assigned to particular categories or classes. Two types of classification are usually recognized: supervised and unsupervised. In the former the classes are defined *a priori* by the interpreter whereas in the second the classes are derived *a posteriori* as a result of analysis.

Sections 4.2 to 4.5 of this chapter deal with various methods of image enhancement, Sections 4.6 and 4.7 outline the principles of human interpretation, and Sections 4.8 onwards are concerned with machine assisted classification.

It is important conceptually to distinguish these three basic types of procedures (preprocessing, enhancement and classification) from the physical modifications and operations which are actually carried out on the imagery, since the same operation may perform very different functions. For example, obtaining the ratio of the values of two spectral bands from a multispectral sensor (i.e. ratioing) has been successfully used both for preprocessing, to help eliminate atmospheric distortions (e.g. Malila *et al.* 1975), and as a method of image enhancement to enhance alluvial features (Merifield & Lamar 1975). Similarly simplification of images by representing tonal ranges with single tones or colours (i.e. density slicing) can be used as a method of enhancing subtle tone contrasts within an image (Ranz & Schneider 1971), or for simple classification (Frazee *et al.* 1972a, b).

4.1.3 *Input and output systems*

Traditionally, the photo-interpreter has required relatively little external aid for interpretation. Many techniques, although conceptually simple and apparently potentially useful, may require sophisticated input and output devices.

The need for these arises in part from the different forms in which imagery can exist and in particular due to the distinction between *analogue* and *digital* forms. In the former the data are represented proportionally by a continuous signal with no discrete values whereas in the latter they are represented by discrete signals with a fixed interval between adjacent values. Examples of analogue signals include most direct pictorial representations of remote sensing data such as photographs and television pictures. Most computers operate in a digital mode and because of this analogue-to-digital converters are necessary if digital methods are to be applied to the products of photographic sensors. Analysis of photographic imagery has often been preceded by digitization using a microdensitometer which relies on variations in the transmission of a small light beam which is shone through a transparency (e.g. Owen-Jones 1977, Williams 1975) though faster flying-spot scanners are increasingly more commonly used. Television cameras, with an analogue-to-digital converter are also frequently used as input devices.

The need for output devices arises because non-photographic sensors often provide data in a digital form, which usually needs to be transformed into hard-copy pictorial form, as may the results of image processing and classification from digital analyses. Careful selection of overprinted characters allows conventional line printers to produce pictorial output (Gonzalez & Wintz 1977), but have a resolution too coarse for many purposes. More suitable devices are available which can create images with resolutions similar to conventional photographs. Some of these, such as laser beam recorders, are sophisticated and expensive but more modest optical-mechanical equipment which images directly on to film (Steiner & Salerno 1975), such as the Optronics Photowrite, can produce very satisfactory results.

Output from video systems is commonly displayed on television monitors which have only moderate resolution compared with photographs. Experience using such devices suggests that only a relatively short period is necessary to accustom the interpreter to their use. The cost of input and output devices can hinder their adoption and this restricts use of new methods of enhancement and classification.

4.1.4 *Image processing as an aid to interpretation*

Processing methods are broadly divisible into *point* and *spatial* processing. In the former, transformations of any individual part of the image depend only on the properties of the point itself (Steiner & Salerno 1975). Included in this category are various methods of combining separate images, such as those from different spectral bands. In contrast, spatial processing of an image involves the properties of the individual point and its surrounding points. It includes therefore the study of 'texture'.

Use will be made of the term picture element or *pixel* as the smallest component of images which is subjected to processing. The size of pixels may simply be that of the original pixels from linescanner imagery (Fig. 2.2): as derived from photographs it is more arbitrary, determined for example by the diameter of the beam in a microdensitometer.

Often several different techniques can be used to implement the same image transformations; for example, density slicing can be carried out by photographic techniques, video-processing methods or by digital computer. Choice of the appropriate method will depend in part upon the local availability of resources and skills. It would be an error to make the assumption that little or nothing can be done without heavy expenditure on special purpose equipment. However, digital methods are gaining in importance. Useful collections of papers may be found in Bernstein (1978) and Aggarwal *et al.* (1977), devoted to digital image processing.

4.2 Processing of tone (point values)

4.2.1 *Analysis of tones on single images*

Images often contain so much data that their simplification can assist interpretation. The most common form of simplification is density slicing which has proved valuable in a wide variety of situations (Ranz & Schneider 1971, Tapper & Pease 1973, Rohde & Olson 1970, Smedes 1971). Density slicing involves the objective identification of areas of equal tone or density within an image. The results can be displayed either by representing a range of tones by a single tone or colour or by drawing contours along lines of equal density. Plate 2 shows the result of two alternative slices of the same image. In the first, the whole tonal range of the scene has been subdivided into eight equal parts. Broad slices of this type are especially useful for making objective comparisons of tone between noncontiguous areas. This is important since the human eye's perception of the darkness of a tone is often influenced by surrounding tones. In the second example in Plate 2 a smaller part of the tonal range has been sliced, showing the usefulness of this technique in analyzing areas with tones too subtle for the eye properly to perceive.

61

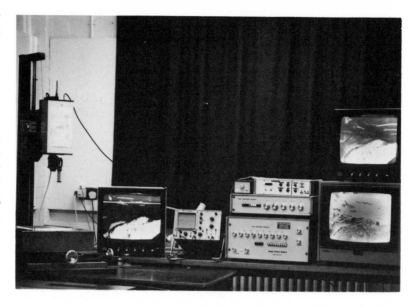

Figure 4.1 A video-processing system. Input is provided by the television camera on the left with the transparency illuminated from below. The analyst can directly control the degree of contrast stretching, density slicing and edge enhancement. Various types of processed images can be simultaneously displayed on the three monitors.

Examples of the application of density slicing techniques include location of soil boundaries in rangeland in South Dakota (Frazee *et al.* 1972a, b) using near-infrared photography and thermal infrared imagery; up-dating of maps of oak woodland in southern Italy (Justice *et al.* 1976); and location of off-shore cold fresh water springs in Hawaii (Ranz & Schneider 1971). Care must be taken not to make excessively distant extrapolations of ground conditions simply on the basis of tone (Evans *et al.* 1976), but, if there is sensible use of density slicing by Earth scientists who are aware of the basic terrain units found within an area (Justice *et al.* 1976), the technique can form a simple but very effective extension of the interpreter's skills (Verstappen 1977).

Production of the density slices illustrated in Plate 2 was executed using a video processing unit relying on a television camera for input and a television monitor for output (Fig. 4.1), similar to that described by Schlosser (1974) and Barnett and Williams (1979). Density slicing can also be carried out photographically using Agfa contour film (Ranz & Schneider 1971, Nielsen 1974). This method can produce very high resolution results but is not directly in the hands of the interpreter and is relatively time consuming. Density slicing is a simple operation on a digital computer assuming that input is available in a suitable form. Images may frequently be improved in appearance by contrast stretching (Fig. 4.2), that is transforming the image so that a better range of tones between black and white is used. Assigning equal frequencies of pixels to each part of the range (i.e. histogram equalization) often improves the appearance of images (Donker & Mulder 1976), but even better results can be obtained by directly specifying the form of the histogram interactively (Gonzalez & Wintz 1977).

4.2.2 *Processing of multiple images – addition, subtracting and ratioing*

Multiple images of the same scene increase the complication of interpretation but also offer much greater possibilities for point processing than single images. By combining the original images in various ways a large number of secondary derivative images can be produced. Such combination of images can serve two basic purposes: firstly, the imagery is more manageable because of the reduction in

(a)

(b)

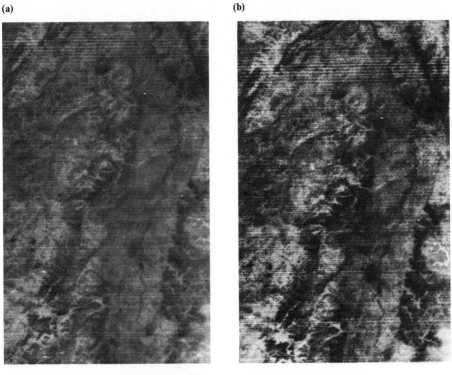

(c)

Figure 4.2 Effects of contrast stretching by photographic means a Heat Capacity Mapping Mission (HCMM) image of the Rhine Valley between Mainz and Freiburg. In this night-time image darker tones represent higher temperatures, and thus the Rhine appears darker than the surrounding land because of its higher temperatures. The margins of the down-faulted alluvial valley floor of the Rhine can be readily detected. Within the flood plain the darker tones and hence higher temperatures of the circular urban areas are apparent. Clearly the amount of information which is extractable by the human interpreter is much greater for the two images with greater contrast. Area depicted: 110 × 250 km. NASA image A-A0034-02130-3, 30 May 1978.

number of images; secondly, judicious combinations of images can lead to many features being revealed that otherwise would be difficult or impossible to detect.

A direct extension of density slicing described in the previous section is to produce two- or three-dimensional slices (Barnett & Willams 1979). Subtraction of one

63

image from another of the same scene means that variations in tone of the resultant image will indicate differences between the two images. This gives the potential for detecting change if the two images are taken at different times as carried out, for example, by Poulton (1975) in rangeland change detection. In practice, many of the differences between two images will be a result of variations in Sun angle and orientation and overall illumination levels so that attempts must be made to eliminate them and overall differences if changes in land cover are to be detected (see also Section 4.14).

Methods for subtracting images include making a print from a combined negative and positive transparency, though this will only show the magnitude of change and not whether it is positive or negative. Digital methods are, of course, readily applicable to the subtraction of images and Humiston and Tisdale (1973) describe a system for change detection with automatic scale correction and registration. Difference-pictures have been used by Goetz *et al.* (1975, p.115) in geological studies in Arizona and were found, somewhat surprisingly, to be almost as useful as ratioed images, which are described below.

Division of corresponding tone or density values of one image by those of another is commonly called ratioing and has proved one of the most useful methods of simple image processing (Sabins 1978). This is because it often enhances subtle contrasts difficult to detect on the original images, especially when channels from different parts of the spectrum obtained simultaneously are derived and used to produce a secondary image. Merifield and Lamar (1975) show how the appearance of alluvial fans near Mojave, California on Landsat multispectral scanner imagery is very much enhanced by ratioing channels 5 and 7. Other applications of Landsat data stress the need for choosing the appropriate ratio: Dethier (1974) found that

Figure 4.3 Processing of multiple images: effects of ratioing Landsat images of Calabria, southern Italy. (a) Landsat image, channel 5; (b) Landsat image, channel 7; (c) ratio of channels 5 and 7. Relief effects have been much reduced and land cover types thereby enhanced. This is especially noticeable for areas of woodland in the north of the area depicted by the darkest tones. Note that for areas actually in shadow (that is receiving no direct illumination only diffuse irradiance), ratioing does not apparently reduce the topographic effect. Area depicted: 30 × 30 km.

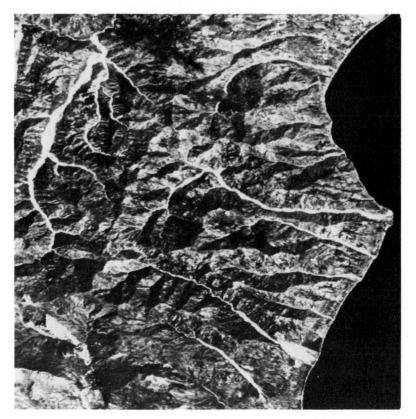

(a)

(c)

the 6/5 ratio was most useful for biomass studies in an America-wide study of phenological changes; for geological and pedological applications, Siegal and Goetz (1977) found that the effects of vegetation were minimized by using ratios 4/5 and 6/7 in a study of soil conditions using multispectral scanner data from aircraft of part of south-eastern Pennsylvania. Wagner *et al.* (1973) found ratioed images, both from within the visible and thermal parts, useful in detecting surface material differences and soil moisture conditions. Vincent (1975) has demonstrated how ratioing different parts of the thermal part of the spectrum results in images in which temperature variations are suppressed and chemical and mineral variations in silicates emphasized. One of the results of ratioing can be to reduce the effects of unequal slope illumination caused by varying slope angles and aspect (Vincent 1972, 1973; Holben & Justice 1980), but will not eliminate shadows (i.e. areas illuminated only by skylight) and Landsat six line banding will be enhanced.

Figure 4.3 shows the effects of ratioing two images in which the effects of relief are clearly reduced and the surface detail is enhanced revealing the land cover pattern much more clearly.

More complex ratios have been used, in particular ratios of the form $(X_1 - X_2)/(X_1 + X_2)$ by Shrumpf (1975) and Justice (1978): subdivision of one channel by the sum of others $(X_i/\Sigma X)$ is commonly called normalization.

The only effective method for the production of ratioed images is by digital means, though the ammonia diazo process described below can be used with some difficulty (Malan 1976).

The other two basic arithmetical manipulations of addition and multiplication have been used more infrequently in the production of monochromatic

Figure 4.4 Stereoscopic image of central Nepal obtained from adjacent Landsat passes. A pocket stereoscope is required to obtain the three dimensional image. Area depicted: 80 × 135 km. NASA images: E-1146-04283-7 and E-1147-04341-7.

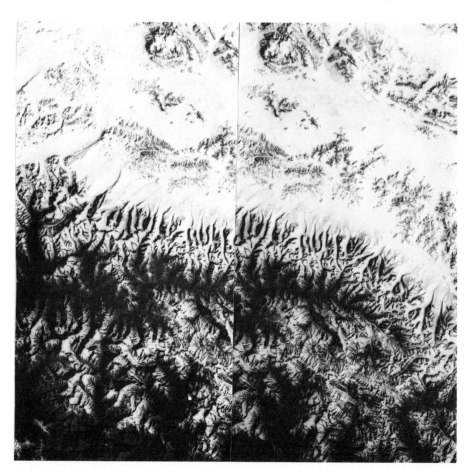

66

images though *colour* additive viewing has proved very useful. Multiplication of images tends to enhance the ruggedness of terrain. Transformation of the original images by deriving their principal components has been used to enhance imagery. These components are linear combinations of the original bands, weighted by coefficients in order that the first axis passes through the direction of maximum variation in the feature space (see Section 4.8), the second at right angles through the direction of maximum residual direction and so on. This method has been found useful for enhancing relief (Donker & Meijerink 1977) and for the study of land cover (Meijerink & Donker 1978).

Before consideration of colour additive viewing, it is appropriate here to include the consideration of stereoscopy which is a traditional form of image processing for combining images, though not normally regarded in this way. By viewing the two images of the same scene taken from different angles, normally using a stereoscope, the perception of a solid image is possible for most people thus permitting direct examination of the relationship between spatial patterns and relief (e.g. Figs 5.5 and 5.6). Stereoscopy apparently is unique in its dependence on the human brain for carrying out the image processing directly. The theory and practice of stereoscopy has been well documented and the reader is referred to standard texts for further information (American Society of Photogrammetry 1960, von Bandat 1962). Although normally obtained using conventional black and white air photographs, stereoscopy can be taken advantage of with many of the other types of remotely sensed imagery. The use of sets of radar imagery taken from different flight paths has been found to have substantial advantages over single strip imagery (Koopmans 1974, Poulton 1973) though cross-flight distortions of the imagery can hinder obtaining a fused image.

Although imagery from space altitudes generally gives a poorer stereoscopic image because of the large vertical distance from the ground to sensor relative to the ground's relief, useful results have been reported using Landsat and Skylab imagery (Poulton 1973, van Genderen 1974) and the stereo effect from the overlapping Landsat passes from Nepal can be seen clearly in Figure 4.4. Various plans have been formulated to enable stereoscopic images to be produced from space altitudes for more extensive areas than at present, including a specific satellite christened Stereosat. Even where overlapping images do not exist, Batson *et al.* (1976) have shown how stereo images can be generated by the use of a pair of images one of which has a parallax introduced automatically using digital terrain elevation data obtained terrestrially. Viewing of pairs of different images obtained from exactly the same geographical position has been carried out successfully in producing pseudo-stereoscopic images using vertically and horizontally polarized radar images (Newton 1973).

4.2.3 *Additive colour processing*

Additive colour processing is probably the most common method of processing multiple images, both because of the usefulness of the results and the simplicity of its implementation. The method involves reproducing the variations in picture point values within each separate image by a different colour and then combining and registering the separate coloured images to produce a colour composite (Yost 1972). The images used may be negatives as well as positives, or combinations of both. Use of the latter may produce very dramatic visual results, though interpretation may be more difficult than in the case of addition of imagery of only one type. Plate 3 shows

67

the colour composite prepared from the images in Figure 2.5. It is immediately obvious that the data as represented in the single colour image are much easier to interpret than in the three separate images. This arises in part because of the human eye's ability to discriminate colours much more effectively than grey tones: according to Slater (1975) by as much as 100 times.

Conventionally the colours chosen when combining three images are the primaries, blue, red and green, or their complementaries, yellow, cyan and magenta. Combining more than three images will require the use of further colours, but the resultant composite may add little because perception by the human eye is the result of stimuli caused by mixtures of the three primaries (American Society of Photogrammetry 1968). Where there are more than three images to be combined, it is therefore normally advisable to use some form of preliminary manipulation such as ratioing to reduce the number of images to three.

Colour composites have been used for a wide variety of interpretation tasks especially in the analysis of Landsat imagery. These include its use for geological mapping in arid areas of Colorado (Goetz *et al.* 1975); for forest discrimination in British Columbia (Lee *et al.* 1974); monitoring rangeland changes in central coast mountains of California (Poulton 1975). Poulton (1975) shows how the combination of a magenta positive image of one date with a cyan negative image of a second, allows one readily to monitor changes between the two dates. In 'Landsat views of the Earth' over 350 colour composites of a very wide variety of landscapes from throughout the world are presented with brief interpretations of each scene (Short *et al.* 1976). Detection of flooded areas using the appearance of water-stressed plants in colour composites derived from multiband photography taken from aircraft is described by Meyer and Welch (1975). Peterson *et al.* (1969) have shown how horizontally and cross-polarized radar images can be combined very effectively for discrimination between a variety of natural vegetation communities.

More complicated colour composites can be produced by combining images which have already undergone transformations. For example, ratioed images have been combined to bring out subtle contrasts in surface conditions for the purposes of geological studies (Goetz *et al.* 1975).

Several alternative methods of obtaining colour composites exist. The most direct system involves the use of a purpose-built additive viewer, where each of the separate images have a beam of different coloured light projected through them and are then optically combined (Fig. 4.5). Additive viewers of varying complexity are available, some of which are specifically designed to accept exposed film from multiband cameras such as the I²S Mini-Addcol viewer (Steiner & Salerno 1975, p. 713). A useful form of additive viewer is described by Beaumont (1977) which consists basically of three colour photographic enlargers with the optics adjusted to give a registered image. This system has the advantage of allowing production of a large colour composite image as well as allowing direct production of a hard-copy photographic print. Photographic production of colour composites usually involves the use of a single enlarger, exposing each image separately on to colour film using a different colour filter (or producing dyed images), registering them and exposing this composite on to film.

A very popular alternative to photographic processing is the use of diazo processing (Fig. 4.6) which has the advantages over photographic processing of being cheaper and the final result being far more immediately under the control of the interpreter. An ultraviolet radiation source is used to illuminate a transparency of each image in turn, which has ultraviolet sensitive film in contact behind it. Upon exposure using ammonia fumes, a positive reproduction of the image is produced in

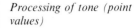

Figure 4.5 Fairey surveys additive viewer. The lamps on the left illuminate transparencies through filters and the registered image is displayed on the screen in the horizontal table surface which also contains controls for adjusting filter colour and illumination levels.

Figure 4.6 Diazo processing system. A transparency is placed in the lower UV exposure unit underlain by diazo transparency paper. Development is then by ammonia fumes in the upper unit.

one of several colours according to the type of diazo film used. Combining three of these images using a different colour film for each image produces a colour composite which can be viewed directly on a light box or projected on to a screen using an overhead projector. A disadvantage of the diazo system compared with photographic processing is that although there is little deterioration in resolution, the full tonal range of photographic film cannot be represented on diazo film. Consequently variation in the lightest and darkest tones may well be lost.

By using different colours to represent the images many alternative colour composites can be produced. Work by Lee *et al.* (1974) shows that the choice of colour combinations is important in different identification tasks (Table 4.1), though certain composites were clearly satisfactory for several tasks. Skaley *et al.* (1977) have developed a colour prediction model for use with diazo film to maximize the colour contrasts of points within composites. More sophisticated digital systems exist which, given digital input from three separate bands, can optimize the colour contrast between any two objects by, for example, ròtation of the three red, green and blue axes (Algazi 1973); these are closely related to digital principle component methods (Taylor 1974, Donker & Mulder 1976) (Plate 5). **69**

Table 4.1 Analyses for the detection of different features using diazo colour enhancement. The numbers refer to Landsat bands (see Table 2.3) and the letters to colours: R (red), B (blue), G (green).

No.	Features	Essential bands and colour to provide the best rating
1	Lakes	6R, 7R, 7G or 7B with 4R or 5R.
2	Hardwoods	6B with 7R.
3	Conifers	5R, 5G or 5B in composite of 3 bands
4	Water pipe lines	4B5G7R
5	Dams	4R5G
6	Logged areas	4G6B, 4R7G, 4G5B7R, 4G7R, 5R6B7G or 5R7G
7	Logged and burned areas	4R7G, 4G5B7R, 4G7R, 4R5B7G, 5G7R, 5R6B7G or 4B5G7R
8	Log booms	4R5G or 4B5G7R
9	Logging roads	4G5R, 5G6R, 4G5R7B or 4R5G
10	Highways	4B5G or 4R5G
11	Railways	4B5G
12	Power lines	4R5G

4.3 Spatial processing

4.3.1 *Introduction*

Human interpreters commonly use the spatial arrangements of tones for interpretation and in their analysis of conventional monochromatic air photographs they are often more important than tone. Both 'texture' and 'pattern' are used by image interpreters to describe spatial variations but are not normally regarded as synonyms. No hard and fast definition of either exists but the term 'texture' is normally used where the individual parts of an area of consistent spatial variation are too small or indistinct to be recognized separately: in contrast an image or photo-pattern is normally regarded as an assemblage of tone or texture areas which have a distinctive spatial arrangement, each of which can be separately recognized and potentially classified. The difference between texture and pattern is, therefore, partly a function of the scale of the imagery relative to the ground surface (Carter & Stone 1974) and a 'texture' at one scale may appear as a 'pattern' at a larger scale. Some of these relationships are illustrated in Figure 4.7, of the same piece of ground viewed at a variety of scales. A similar relationship exists between tone and texture, since a texture on aircraft imagery may well appear as a tone on satellite imagery.

The use of the terms texture and pattern as described above corresponds only approximately to those found in the quantitative study of image properties found in the pattern recognition literature (e.g. Haralick 1973, Rosenfeld 1976). Texture is often taken to refer to virtually any type of spatial variation (Thomas 1977) and thus subsumes a large part of the above definition of 'pattern'. However, it does not include contextual information; i.e. information about the recognized identity of the parts of the image which aids identification of other parts (Section 4.13). Lastly, and confusingly for the traditional interpreter, we note that in the subject of pattern recognition, a pattern is not necessarily characterized by spatial properties, since we can refer to the pattern of a single pixel using the tonal values in several spectral bands to describe it. Thus a pattern is any well defined set of measurements (Swain 1978) or can be defined as a 'quantitative or structural description of an object'

(a)

(b)

Figure 4.7 Effects of scale on definition of tones, texture and patterns in human interpretation terms. Note how areas of tone on one image become textures or even patterns at large scales. Landsat image (E-2189-10224-Channel 5) of south central England centred on Greenham Common air base and extending in the upper image from the Severn estuary to west London. Area depicted on first image: 175 × 130 km.

(c)

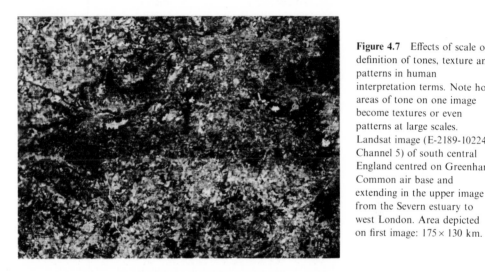

(Gonzalez & Thomason 1978). In the remainder of this section we describe the methods used to characterize spatial variations and in particular those which either lead to improvements in their display to aid human interpreters or which can be used to provide quantitative descriptions useful in automated classification. Because of our interest in these newer techniques we shall eschew the traditional photo-interpreter's use of the terms 'texture' and 'pattern' and adopt those of workers in pattern recognition studies, unless otherwise explicitly stated.

The number of ways of describing the spatial variations in tone within a scene either quantitatively or qualitatively is effectively limited only by man's imagination, like any other set of morphological properties (Tobler 1969). The important tasks are therefore not concerned with simply inventing new descriptions but identifying ones which permit sensible discrimination between significant types of land or ground targets and in identifying measures whose derivations are not too time consuming. Discussion will be limited to those properties whose usefulness is already proven or is likely to be.

Consideration of line and shape recognition is delayed until the section on image classification in Section 4.12. Although line recognition is clearly closely related to the study of lineations discussed in the next section, the latter involves recognition of alignments within the image rather than the identification of lines as features in their own right and their separation from the background. The difference is a subtle one and highlights the close relationship of image enhancement to classification.

4.3.2 *Lineations in imagery*

The detection and recording of linears or lineations on imagery has long been carried out especially in geological studies (e.g. Peterson 1979). Figure 4.9 shows the result of abstracting all perceived lineations from a photograph (Fig. 4.8): the resultant linear map is then used as an aid to interpretation. Lineations may be either boundaries between areas of different tones or lines of pixels with a distinctive tone.

There are several methods which can enhance the human interpreter's ability in analyzing lineations. For example, we can transform images so that changes in image value are displayed instead of the absolute values. We can implement this readily using a video input and analogue processor similar to that described for density slicing. The video signal is transformed such that the output signal is made proportional to the changes in the video signal across the television picture line. Alternatively the gradients between adjacent image points can be calculated digitally and the resultant values used to create an edge-enhanced image (Fig. 4.10). A similar digital implementation is described by Bodechtel and Kritikos (1971).

Application of digital techniques allows edge enhancement in all orientations simultaneously. For example, this may be carried out by application of digital filters to arrays of image values for example using the following matrix (Steiner & Salerno 1975, p. 701):

$$\begin{matrix} 0 & -1 & 0 \\ -1 & +4 & -1 \\ 0 & -1 & 0 \end{matrix}$$

Examples of the application of various non-linear filters are outlined by Rosenfeld (1969).

Whatever enhancement method is used, there remains the problem of subjectivity

Figure 4.8 A black and white rendition of a true colour photograph taken from a Skylark Earth resources rocket of central Argentina, south-west of Cordoba. This was used to produce the linear map in Figure 4.9.

Figure 4.9 Interpretation of linears of area shown in Figure 4.8. Careful inspection of Figure 4.8 will reveal many more linears than are immediately apparent. However many of them are only visible on the original diapositives.

0   5 km

involved in locating and describing the lineations and consequently of reproducibility (Burns *et al.* 1977). Furthermore, these methods are somewhat restricted in that they concentrate on edges or lines rather than the overall orientation of the image. The latter can be more readily studied by the use of Fourier techniques in image analysis (Duda & Hart 1973). Fourier methods can be used to model the variation in spatial (or linear) properties in terms of a series of sine and cosine curves. The Fourier transform $F(u,v)$ of a picture ($f(x,y)$) is defined by,

$$F(u,v) = \int\limits_{-\infty}^{\infty} \int \exp\left[-2\pi\sqrt{-1}\,(ux+vy)\right] f(x,y)\,\mathrm{d}x\,\mathrm{d}y. \qquad (4.1)$$

where u and v are spatial frequencies.

73

The angular distribution of values in $|F|^2$ is sensitive to the directionality of the texture in f (Weszka *et al.* 1976). This equation can be solved either digitally or optically. Although digital solutions have been obtained in studies of terrain texture by use of the discrete Fourier transform (e.g. Gramenopoulos 1974, Weszka *et al.* 1976, Hornung & Smith 1973), these are relatively time consuming for image analysis even on large digital computers. Therefore, more attention has been paid to the use of optical methods based on the ability of converging lenses in performing two-dimensional Fourier transformations since the latter are closely related to the lens' Fraunhofer diffraction pattern (Goodman 1968). Figure 4.11 shows the general arrangement of the most commonly used system. Light from a coherent light source (laser) illuminates a transparency of the image encased in a liquid gate to remove film thickness variation; the resultant diffraction pattern is produced by the Fourier transform lens. Although in theory any convergent lens can be used, in practice specially made lenses are needed for good results. The resultant diffraction pattern obtained from a radar image is shown in Figure 4.12 and overall the alignments of the image are now summarized by its azimuthal arrangement. The sharply defined vertical gap is a result of radar's poor depiction of orientations at right angles to the direction of motion (Harnett *et al.* 1978). Study of differences between these arrangements for different terrain and land cover types has been carried out by several workers (e.g. Gramenopoulos 1974, Arsenault *et al.* 1974, Barnett & Harnett 1977, Bauer *et al.* 1967, Pincus 1969). Azimuthal properties can be used to provide quantitative measures of the directionality of the texture. An automated system has been suggested by Jensen (1973) using a photodetector array of 32 wedge-shaped and 32 annular ring elements on to which the diffraction pattern is directly imaged.

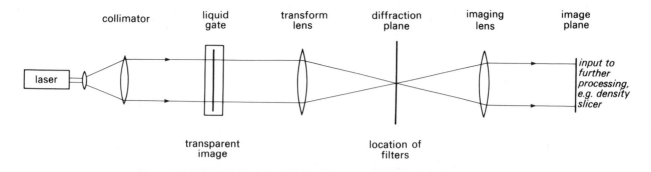

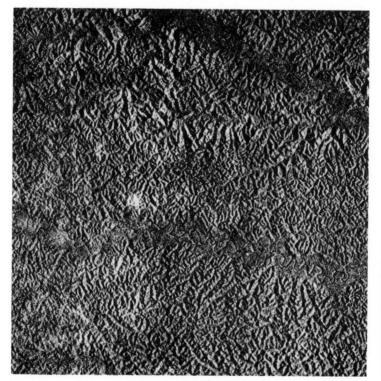

Figure 4.11 Layout of a system for optical Fourier analysis.

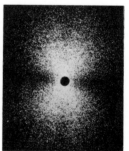

Although potentially of value, the use of these diffraction patterns is limited: a relatively large part of the image has to be analyzed and thus the effective resolution of any resultant terrain map produced in this way will be very much coarser than the original image.

Of apparently greater value to the interpreter is the use of directional filtering. If we place a suitable filter (Fig. 4.13) in the diffraction pattern focal plane of the Fourier transform lens we can then use a further Fourier transform lens to reconstitute an image in which certain orientations have been selectively excluded by the filter (Fig. 4.11). This technique has proved valuable in detecting hidden terrain alignments obscured by other more dominant trends (Verstappen 1974, 1977, Harnett *et al.* 1978). Figure 4.14 shows the result of using a narrow bow-tie filter in which the dominant line orientation has been removed and the underlying patterns more clearly shown. Since the system is optical, the transformations occur in real time and the interpreter has the potential for selecting different orientations by rotating the filter, altering widths of cut-offs of the filter and seeing the immediate result of the changes.

Figure 4.12 A radar image of the Upper Amazon Basin and its diffraction pattern (produced by Michael Barnett and Peter Harnett, Department of Physics, Imperial College, London). Note the marked absence of orientations normal to the flight line caused by the functioning of the Radar system (see Ch. 2). Area depicted: approximately 75 × 75 km.

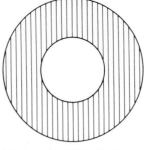

(a) Bow-tie filter for directional filtering

(b) Annular filter for low-pass filtering – excludes high spatial frequencies

(c) Circular filter for high spatial frequencies – excludes low spatial frequencies

Figure 4.13 Types of filter for directional and spatial filtering.

Figure 4.14 Results of applying a directional filter to a Landsat scene of part of Botswana, north of the Makarikari salt pans (produced by Michael Barnett and Peter Harnett, Department of Physics, Imperial College, London). (a) Before filtering. The dominant trends present are the WSW–ENE trending vegetated sand dunes. In the east of the area depicted, an approximately NE–SW trending strip can be seen which represents a monoclinal structure (Short 1976). (b) After filtering. The dominant sand dune trend has been removed revealing underlying patterns ↓ more clearly. For example, inspection of the eastern part of the image reveals several lineations parallel to the trend of the monoclinal structure, which are much less apparent in the original. Area depicted: 66 × 88 km.

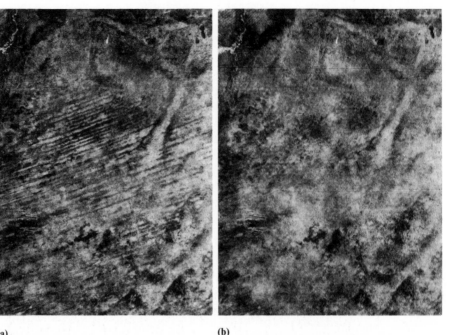

(a)

(b)

4.3.3 *Spatial regularities in imagery*

Use of the Fourier transform for the study of directionality, leads us naturally into a consideration of *spatial frequency*. The most common descriptor by human interpreters of texture is usually with reference to its coarseness or fineness. Often we may perceive that there is more than one level of coarseness: a fine texture may be superimposed upon a coarser one. Unravelling several different trends may prove beyond the discerning power of human interpreters. The radial distribution of values of $|F|^2$ (in equation (4.1)) is sensitive to texture coarseness. High values of $|F|^2$ concentrated near the origin or equivalently bright areas near the centre of the Fraunhofer diffraction pattern represent coarse textural variation, and higher frequency of finer textural variations are found progressively away from the origin or centre. Examination of spatial frequency in this way has been carried out, for example, by Gramenopoulos (1974) and Barnett and Harnett (1977). Jensen's method of rapidly quantifying diffraction patterns described above is also applicable (Jensen 1973), and digital analyses have been carried out by Weszka *et al.* (1976), Gramenopoulos (1974), Ulaby and McNaughton (1975) and Stanley (1978). The qualitative or quantitative analysis of spatial frequency in this way suffers from the enlargement of effective resolution cells as was the case for directional analysis.

Thus *spatial filtering* which is the equivalent of directional filtering seems likely to prove more useful.

Optical spatial frequency filters are comprised of annular forms (Barnett & Harnett 1977). In Figure 4.13b the low frequencies are blocked by a central circular disc allowing only the high frequencies to pass, and conversely Figure 4.13c shows a low-pass filter. Filters intermediate between these two could be produced to allow any selected frequency through.

Figure 4.15 shows the effect of a high-pass filter on the diffraction pattern in Figure 4.12. The high spatial frequencies of the dissected areas are passed by the filter causing these areas to be bright, whereas the flat alluvial valleys composed of low spatial frequencies appear dark. These textural regions have been converted into tonal ones and the differences enhanced (Harnett *et al.* 1978). Once again one of the prime virtues of the system is that it is interactive in allowing the interpreter to perform sophisticated transformations, immediately to see the result, and rapidly to modify the transformation whilst carrying out his interpretation.

A fundamental assumption of the Fourier transform approach is that it assumes a picture is periodic even if it is not (Weszka *et al.* 1976). For example, it may contain many very abrupt transitions and thus consequently it will be somewhat unsatisfactorily modelled as the sums of sine and cosine curves. One possible solution, which has been suggested, is use of the Hadamard transformation in which modelling is by sums of rectangular shapes (Steiner & Salerno 1975, Morgan 1977). However, optical solutions are not possible and the number of operations necessary can be very large, rising proportionately as fast as for Fourier transformation (namely by $n \log n$, where n is the number of pixels).

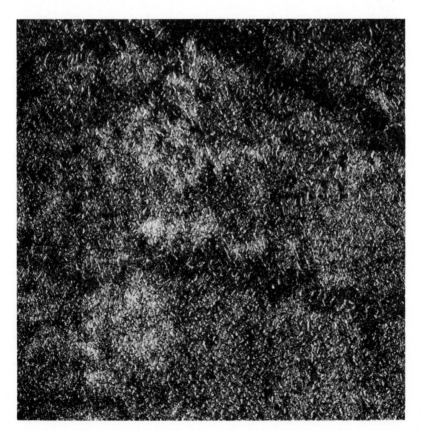

Figure 4.15 Results of applying a high-pass filter to the SLAR image in Figure 4.12 (see text for explanation).

77

Spatial and directional filtering can also be carried out digitally. Goetz *et al.* (1975) have applied a variety of digital filters to arrays of Landsat data; the resultant transformations are then reproduced as high quality images for visual interpretation.

An alternative approach to the study of texture involves the use of statistical properties of the scene's picture points. For example, if we were to subdivide a scene into regular cells, we could calculate for the picture points of each of these subsamples, simple statistics such as the ranges variance, skewness or kurtosis. For example high variance-values would then indicate high variability and low values low overall variability.

A rather more complex set of first-order statistics was suggested by Weszka *et al.* (1976), based on grey level differences. For a picture with m different grey levels we may define a probability density p_δ whose m vectors each represent the probability that a given difference i between two picture points will occur for a displacement δ. On this basis it is possible to calculate various measures, e.g.

$$\sum_{i=1}^{m} i^2 p_\delta(i) \quad \text{and} \quad \frac{1}{m} \sum_{i=1}^{m} i \, p_\delta(i) \tag{4.2}$$

One restriction on the usefulness of these relatively simple first-order statistical properties is that little account is taken of the internal spatial structure within the subsamples.

Haralick *et al.* (1973) proposed a scheme using second-order statistical properties by means of a picture's grey-tone dependence matrices. The cells of this matrix contain the numbers of times any given grey tones i and j are neighbours within a given subregion of the image. The principal diagonal therefore shows the number of times neighbouring image points have the same value and consequently the occurrence of a large proportion of values in this diagonal would indicate a relatively coarse texture. Separate grey-tone matrices can be defined according to the particular direction and nearness of a neighbour. Any given point in a rectangular array has eight neighbours (Fig. 4.16), and Haralick *et al.* (1973) using four pairs of these therefore defined four grey-tone dependence matrices. Differences between these matrices will indicate directionality of the texture. A series of 28 measures using these matrices was also derived, the most useful of which were the angular second moment, contrast, correlation and entropy as defined in Figure 4.16. These were calculated for each of the above four grey-tone dependence matrices and the mean and range calculated. Their usefulness was examined both for aerial photographs and Landsat data from channel 5, each sample being 64×64 pixels in size. The latter were used for the most rigorous testing with a training set of 314 samples being used to predict the character of another 314 samples with an accuracy of 70.5% which is encouraging despite the wide range of land use types which were defined. Cross-band textural measures have also been used but without improvements in classification (Haralick & Bosley 1974).

A useful comparison of the value of various texture measures in discriminating between various terrain types has been carried out by Weszka *et al.* (1976). Their principal comparison was for a Landsat 1 image of eastern Kentucky for terrain underlain by three different rock types. Results indicate that measures based on the grey-tone dependence matrices were better discriminators than features derived from digital Fourier methods. However, interestingly a first-order grey level difference statistic based on $(1/m)\Sigma i p_\delta(i)$ described above proved to be as efficient

(a) Eight nearest neighbours of pixel X, according to angle, θ:

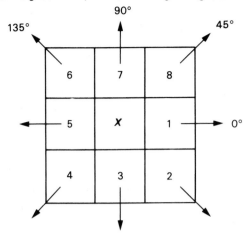

For $\theta = 135°$, 6 and 2 are the nearest neighbours
For $\theta = 0°$, 5 and 1 are the nearest neighbours, etc.
$G = \{1, 2, ..., N_g\}$ is the set of possible pixel values
$P(i, j, d, \theta)$ is the matrix of relative frequencies which 2 neighbouring pixels have with each other, separated by distance d, and angular relationship θ, one with pixel value i and pixel value j

(b) Thus for 4 possible values of θ, we obtain 4 matrices.
Example: for the following picture composed of 4×4 pixels, and only 3 possible pixel values (1, 2 or 3):

1	1	3	3
1	2	2	1
2	2	1	2
1	3	3	3

we obtain the following 4 spatial dependence matrices:

$$P(i, j, 1, 0°) = \begin{matrix} 142 \\ 420 \\ 203 \end{matrix} \qquad P(i, j, 1, 45°) = \begin{matrix} 211 \\ 123 \\ 130 \end{matrix}$$

$$P(i, j, 1, 90°) = \begin{matrix} 152 \\ 513 \\ 230 \end{matrix} \qquad P(i, j, 1, 135°) = \begin{matrix} 042 \\ 412 \\ 220 \end{matrix}$$

(c) Using any of these spatial dependence matrices various texture measures can be derived, e.g.

(1) $T_3 = \Sigma_{i=1}^{N_g} \Sigma_{j=1}^{N_g} \dfrac{(P(i, j, d, \theta))^2}{R}$ (angular second moment)

(2) $T_2 = \Sigma_{i=1}^{N_g} \Sigma_{j=1}^{N_g} \dfrac{P(i, j, d, \theta)}{R} \log P(i, j, d, \theta)$ (entropy)

(3)

$$T_3 = \frac{\Sigma_{i=1}^{N_g} \Sigma_{j=1}^{N_g} (ij\, P(i, j, d, \theta)/R) - \mu_x \mu_y}{\sigma_x \sigma_y}$$

(correlation)

where μ_x and μ_y and σ_x and σ_y are the means and standard deviations of the marginal distributions obtained by summing the rows and columns of $P(i, j, d, \theta)/R$, and R is the number of pixel pairs used in obtaining each matrix.

as the Haralick-type measures, computationally quicker to derive, and thus was recommended. All the methods proved more successful than two interpreters, one of whom was a trained geologist.

The use of statistical properties for describing texture has been primarily for defining features for use in methods of automated classification (Section 4.8). A major difficulty in using many of these techniques is that they require grouping of substantial sets of pixels. Commonly areas of 64×64 pixels have been used when analyzing Landsat data (Haralick & Shanmugam 1974, Haralick & Bosley 1974, Weszka *et al.* 1976), so that a very substantial reduction in the effective 'resolution' of any resultant map will inevitably occur. Small cells could be used, but these would automatically exclude coarser textural variations and also with small sample sizes lead to unstable texture values based simply on the chance location of the grid. It follows that in automated classification systems if high resolution results are necessary we would do well to rely on spectral properties so far as possible and only if the results are not satisfactory should we resort to textural properties. This somewhat gloomy prognosis does not apply to the value of transformed images in which particular textural components are enhanced or suppressed; several of these methods seem likely to prove increasingly valuable in the future.

Figure 4.16 Eight nearest neighbourhood resolution cells and use in deriving textural measures (after Haralick & Bosley 1974).

4.4 Choosing the appropriate image enhancement technique

The very diversity of available methods for enhancing images to aid interpretation, as described in the previous sections, inhibits generalizations concerning their utility. Nevertheless, it is clear that many image enhancement methods have reached a stage in their development where they are either regularly included or could be readily incorporated into operational terrain resource programmes.

A restriction on the usefulness of these methods stems from the study of terrain being pre-eminently a field subject. Although laboratory results can be taken into the field and field information returned to the laboratory, there is a need for image enhancement (and classification as described below) to be carried out in the field itself.

The plethora of methods potentially available poses two related problems for image interpreters. One may ask whether it is worth investing time (and money) in learning how to use these new techniques, recognizing that initially results may be poor until expertise is acquired. Having obtained some expertise, a final evaluation of whether the methods provide better and/or cheaper and/or faster interpretations compared with conventional methods needs to be made. A proper evaluation will often be very complex and difficult to achieve and partly because of this, the new approaches may well be rejected in favour of the traditional unless benefits are especially marked. Thorough and objective testing is often sadly lacking for many methods. This arises in part from a desire not to expose them to the full rigours of objective evaluation, but also because tests which allow the prediction of a technique's capabilities in a variety of conditions and for different purposes are extremely difficult to devise. Development of appropriate evaluation methodologies is a challenge to research workers if a sensible development of more advanced image enhancement methods is to be achieved.

4.5 Elements of interpretation

Image interpretation in the present context is the act of detecting, recognizing and classifying the Earth's surface and subsurface characteristics using remotely sensed data. Whether human or machine-based interpretative systems are being considered, four basic sequential stages can be recognized:

(1) selection of features;
(2) extraction of features from the images;
(3) definition of rules delimiting classes;
(4) assigning parts of images to classes.

We note parenthetically that the latter two procedures are commonly and confusingly both called 'classification'. Human interpreters, as judged by the sophistication of their results, display considerable expertise in carrying out these procedures. We can gain an understanding of the strengths and weaknesses of human image interpretation and the need for automated methods of classification by looking briefly at the different techniques or approaches that have been used in the former.

4.6 Methods of human interpretation

The human interpreter has a variety of methods available for identification and classification of ground surface properties.

Broadly these can be divided into those using keys, analogues, and deductive reasoning; the categories are not completely exclusive and elements of two or more may well be used during any one interpretation exercise.

4.6.1 *Use of keys*

Keys have commonly been used in image interpretation because they can often be applied successfully by relatively inexperienced interpreters, at least for simple tasks. They are often comprised of a set of images against which each part of the image being analyzed is matched and allocated to the class whose image in the key it most resembles (Coiner 1972). Alternatively the key may be composed of a series of step-by-step instructions which eliminates incorrect choices and successively narrows down to the correct identification.

Estes and Simonett (1975) describe a large number of different keys demonstrating the breadth of their possible uses. More complex interpretation than simple identification of land cover type may be achieved by use of *associative keys* which permits deduction of information not directly discernible on images.

The principal drawback of keys is that they are suitable only for relatively 'low-level' interpretation tasks. One might hope often successfully to map land cover by this method, but the accuracy for example of mapping land capability would normally be expected to be low except within very simple landscape types. Keys are also likely to become more complicated as the interpretation task becomes more advanced, and thus become so unwieldy that other approaches have to be adopted (Vink *et al.* 1965, Miller & Schumm 1964).

4.6.2 *Use of analogues*

More advanced interpretation is often carried out by skilled interpreters using analogues, which may well exist only as mental constructs. Such analogues involve a model in which not only are the range and co-relationships of the physical properties understood, but also the various functional relationships are at least partly known. Consequently, exact physical correspondence of part of a scene with any single analogue will be unnecessary for substantially correct interpretation to be made. It is clear that analogues can normally only be successfully used where the interpreter possesses considerable specialist expertise relating to the nature of the feature being observed. Perrin and Mitchell (1969) have produced a very valuable document which can help in the choice of the most appropriate analogue in the study of arid terrain. Terrain types and land units are described in tabular form with, so far as were available, a stereo-pair of air photographs, and ground photographs (Fig. 4.17). In a sense this volume forms an advanced key and might well be called an 'analogous area key' by Estes and Simonett (1975, p. 833), but it is doubtful if any but experienced arid geomorphologists could successfully use this work and any attempt at simple matching would be unlikely to give satisfactory results.

(a)

(b)

Figure 4.17 Example of the
use of analogues for desert
terrain (Perrin & Mitchell
1969). The description
includes: (a) ground
photograph; (b) oblique air
photograph; (c) stereoscopic
pair of air photographs (a
pocket stereoscope is needed
to obtain a three-dimensional
image); (d) block diagram. UK
Crown Copyright: Ministry of
Defence.

4.6.3 *Use of deductive reasoning*

Even the most experienced interpreter will frequently find landscape types for which
he has no satisfactory analogue. Commonly, therefore, it is necessary to make
deductions on the basis of evidence from the image along with ancillary
information. The extent to which one should view individual elements of terrain in
turn or view the terrain as a whole in an integrated way is the subject of some debate.
Estes and Simonett (1975) would argue for the former view. The approach of others,
especially in the use of space-altitude imagery, where identification of many of the
elements of the scene may be impossible because of the small scale of the imagery, is
first to view the patterns on the imagery in a holistic way (e.g. Drennan *et al.* 1974).
Without doubt the latter approach is not for the inexperienced interpreter, but it has
been found useful in the initial subdivision of terrain on space imagery as described in
Chapter 5.

(c)

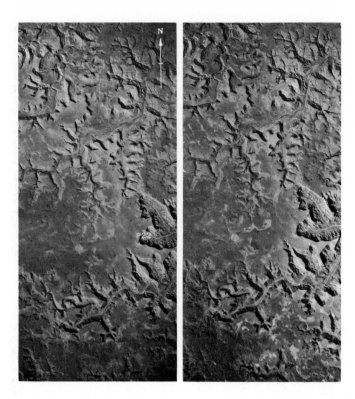

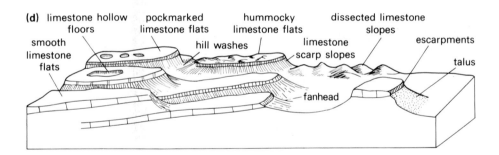

(d) limestone hollow floors · pockmarked limestone flats · hummocky limestone flats · dissected limestone slopes · escarpments · smooth limestone flats · hill washes · limestone scarp slopes · talus · fanhead

Distinguishing features
(1) dissected plateau (which need not be absolutely level) and >40% dissected
(2) consolidated stratified but non-metamorphosed calcareous sediments
    (limestone, chalk, dolomite, marble, calcrete, coral, etc.)

Pattern no. 22: much dissected limestone plateaux

Climatic region: hot arid

Relief: top has little relief (<30 m) but between top and surrounding plains often >30 m;
    >40% dissected

Country rock: consolidated, stratified but non-metamorphosed calcareous sediments
    (limestone, chalk, dolomite, marble, calcrete, coral, etc.)

Associated features: includes the north African "Kem Kem"

Local forms: Alice Springs; Tassili: Mouydir-Ahnet-limestone; escarpment country of
    Algeria-Mali frontier

83

4.7 Physiographic approach to human image interpretation and the potential role of automated methods

As implied above, the type of interpretation methods used will depend upon the complexity of identification and/or mapping tasks which are being attempted. We can extend this idea by recognizing four levels of conceptual complexity in the interpretation task. When interpreting original or enhanced imagery, human interpreters have only two basic types of data available, namely grey-tone or colour variation and the stereoscopic model of relief variation, apart that is from extra-image contextual information (Section 4.13). Interpretation of ground properties follows either directly or indirectly from these (Figs 4.18, 4.19).

All four *primary interpreted properties* depend on both of the directly observable properties, though with varying importance. Thus land cover type will depend mainly on the use of grey-tone or colour variation (including spatial properties as described in Section 4.3), and will also depend to a lesser extent on relief, since the height of vegetation is a valuable clue in determining land cover type. Similarly, land form will depend primarily on the stereoscopic model but this will be enhanced by grey-tone variations, especially those due to shadows. The primary interpreted properties are relatively limited in scope (Fig. 4.18), but the precision with which they can be interpreted is much enhanced by making use of our knowledge of relationships between them.

On this basis we can derive two further levels of interpretation to create a hierarchical model of image interpretation (Fig. 4.19). **Secondary interpreted properties** are surface properties, which are interpreted principally using the primary interpretations along with ancillary information. *Tertiary interpreted properties* are subsurface characteristics derived both from the primary and secondary interpreted properties along with ancillary information and finally inferences about **land suitabilities** will be derived from all three previous levels in a **quaternary stage**. This model of the interpretation process emphasizes its inferential character and hence the care which should be adopted in carrying it out.

The principal interconnections between the properties have not been indicated in Figure 4.19 in the way demonstrated for the first two levels in Figure 4.18, because of the complexity of inter-linkages involved. Amongst the most important are those between surface form and rock type, which have been qualitatively established and refined for many years (e.g. Leuder 1959, Tator 1960, Miller & Miller 1961, Way 1973) and are capable of increasing precision by the additional application of quantitative methods. Additionally, models have been proposed relating slope form to erosion, notably that by Dalrymple *et al.* (1968). Other especially important sets of relationships are those between vegetation and ground water (Zinke 1960, Abrosimov & Kleiner 1973, Vostokova 1973), between vegetation and rock type (Eardley 1941, Tator 1960, Osipova 1973), particularly in the location of mineralized zones by geobotanical anomalies (Brooks 1972, Cole *et al.* 1974) and the relationships of soils, with relief and hydrology (Fridland 1976). Readily observable properties used to make inferences about less accessible ones have been termed 'landscape indicators' by scientists working in the USSR (Viktorov *et al.* 1973).

The model we have proposed (Fig. 4.18) is a simplification in that interpretation at earlier levels can be aided by interpretation at subsequent levels, though the net movement will be strongly in the directions as shown. Implicit in the model is that the origins of land characteristics can at times be inferred and hence conclusions drawn about past or present processes. The success of the interpretations will vary according to the interpreter's knowledge, quality of the images, type of terrain and

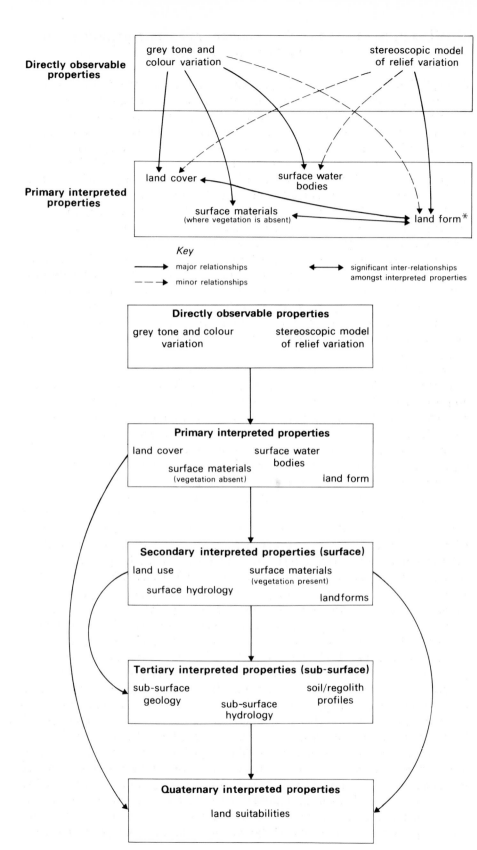

Directly observable properties

grey tone and colour variation

stereoscopic model of relief variation

Primary interpreted properties

land cover

surface water bodies

surface materials
(where vegetation is absent)

land form*

Key

→ major relationships

⟷ significant inter-relationships amongst interpreted properties

--→ minor relationships

Figure 4.18 Relationship between directly observable properties of images and primary interpreted properties (*note* 'land form' and not 'landform' – see text for explanation).

Figure 4.19 Hierarchical structure of human interpretation (note the distinction between land form and landform – see text for explanation).

Directly observable properties

grey tone and colour variation

stereoscopic model of relief variation

Primary interpreted properties

land cover

surface water bodies

surface materials
(vegetation absent)

land form

Secondary interpreted properties (surface)

land use

surface materials
(vegetation present)

surface hydrology

landforms

Tertiary interpreted properties (sub-surface)

sub-surface geology

soil/regolith profiles

sub-surface hydrology

Quaternary interpreted properties

land suitabilities

85

number of stages of interpretation. Ground data collection should always precede, accompany or follow these stages (Ch. 3) and these and other ancillary data sources can be perceived as operating orthogonal to the flows shown.

It should be clear that the potential role of automated classification is by no means equal for each interpretation level and in general will decrease according to the order of the four interpretation levels given above. In the first two, automated methods of image analysis could be the principal technique to be used: in all others they will play a subsidiary role. Production of land use maps has been shown to benefit substantially from machine-assisted classification, but in recognizing origins and functional relationships between various terrain components we would be unwise to expect much direct assistance from it. Nevertheless, the usefulness of automated interpretation should not be dismissed too lightly. Compilation of even morphological properties of the environment can be a time consuming and expensive procedure on which decisions about land planning and land management in the context of regional and national plans will or should be based (N A S 1976). If such descriptions become more readily and cheaply available, there is hope that terrain resources will be used more sensibly and with greater respect. It is from this viewpoint that we approach the study of automated methods.

Automated procedures of image interpretation stem largely from the study of not only remotely sensed imagery but also such topics as character (letter and number) recognition, bubble-chamber scanning, diagnostic cytology and speech recognition (Sklansky 1973, Rosenfeld 1976). Introductions to pattern recognition are provided by Duda and Hart (1973), Steiner (1970), Grasselli (1969) and Tou and Gonzalez (1974). Within remote sensing these methods have been applied primarily to studies of agricultural land use, which involved relatively simple identification tasks compared with those found when making higher level interpretation especially in areas with more complex land covers.

4.8 Feature space and feature selection

In automated recognition systems, it is customary to indicate the similarity between the patterns of pixels (or larger parts of images) by use of a *feature space* (Fig. 4.20), which is defined by axes representing properties or features which are extracted from the image. These properties are usually spectral or spatial image properties. We should recall from Section 4.3.1 that the term pattern in the present context, does not only refer to spatial properties but to any quantitative or structural description.

The number of dimensions of the feature space may become large if several channels have been sensed or if spatial properties are included. Many of these may serve little useful purpose, either because they have inherently low discriminatory power in distinguishing between the categories with which we are interested or because they are redundant since their powers of discrimination are closely matched by other properties. Inclusion of these properties during subsequent classification could increase computing time substantially. A number of computational methods, such as step-wise discriminant analysis and principal components analysis, have therefore been used to eliminate these unwanted properties, the general name of the procedure being *feature selection*. Further details of feature selection methods can be found in Steiner (1970), Fu (1976) and Morgan (1977).

Classification can proceed according to two distinct methodologies. In **supervised classification**, the classes are decided upon *a priori* and the objective becomes one of satisfactorily partitioning the feature space such that members of each class occupy

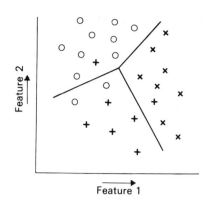

o members of class A

+ members of class B

x members of class C

	Class pixel is assigned to, in feature space			Errors of omission
	A	B	C	
A	9	2	0	2
Class pixel actually belongs to B	1	5	1	2
C	0	0	8	0
Errors of commission	1	2	1	22 = sum of principal diagonal elements

overall accuracy = $(\frac{22}{26} \times 100)\%$ = 84·6%

Figure 4.20 Partitioning of a feature space showing errors of omission and commission in classification.

a unique subspace. In **unsupervised classification** the groupings of patterns are themselves used to define the classes and classification is therefore carried out *a posteriori*. The distinction between the two approaches has been recognized in many other disciplines (Mather 1976, Krumbein & Graybill 1965, Sokal & Sneath 1963) and consequently many of the mathematical methods will be familiar to those who are acquainted with multivariate statistical methods.

The term 'automated' is often used to describe both supervised and unsupervised methods, but strictly 'semi-automated' or 'computer-assisted' are more appropriate since the analyst nearly always has to make important decisions before, after and even during, their execution. Implementation of these methods is usually by software written for digital computers. Two of the most commonly used software packages are LARSYS, developed by the Laboratory for Applications of Remote Sensing (LARS 1967, Swain & Davis 1978) and VICAR created at the Jet Propulsion Laboratory of the California Institute of Technology (Steiner & Salerno 1975, pp. 734–9).

4.9 Supervised approach to classification

Supervised classification requires a **training set** with known ground conditions and image measurements, the members of which are representative of the classes into which the whole of the target area (Section 3.1) is to be assigned. If we can then successfully find subspaces of the feature space which only contain members of single classes, the remaining parts of the target area for which we possess only image properties can be classified. In practice, success is not so readily won. Figure 4.20 illustrates a simple hypothetical example with a two-dimensional feature space and only three classes, which have been separated using linear boundaries. The equations of these lines are of the form:

$$d(X_1, X_2) = C_1 X_1 + C_2 X_2 + C_3 = 0 \qquad (4.3)$$

where the C's are coefficients. Any observation falling on the boundary yields $d(X_1, X_2) = 0$ when substituted in this equation. If positive or negative, the observation will then be assigned to one of the two classes on either side of the line. Notice firstly that groups A and B can only be distinguished using both of the features, but that C could be distinguished from both by using only one of the features. Secondly, note that the partition of this space is realistic, since the lines do not perfectly separate the groups. Two types of error can occur, namely those of omission and commission. Thus groups A and B all suffer from **two errors of omission** since a member of each group is found within the subspace of other groups, but the **errors of commission** are one for group C, one for group A and two for group B, representing the number of outside members wrongly assigned to each of these groups. Judgement about whether all errors are equally important or not will affect the location of the boundaries using the methods described below.

The boundaries need not be linear though non-linear ones will usually be computationally more complex and possibly not worthwhile. The usefulness of the boundaries should be assessed by means of a **testing set**, for which ground conditions are known, in order to assess the more general applicability of the decision rules. If they are found to be satisfactory extrapolation can then be made to the target population, using the spectral data alone, either by substitution of values in the above equation or using look-up tables (Shlien & Smith 1975).

A variety of methods exist for deciding on the location of the boundaries (Steiner 1970). If we are dealing with only two dimensions, boundaries could be drawn relatively easily and often with satisfactory efficiency by eye. A simple numerical method would be to construct lines at right angles mid-way between lines joining group centroids (or in other words apply the minimum distance rule). Alternatively, deterministic boundaries composed of square or hexagonal shapes can be defined by appropriate algorithms (Morgan 1977). A more complex approach is to use probabalistic methods. The most common method is to assume that the training sample of each class has a multivariate normal (Gaussian) probability density distribution so that the location of each can be defined by the position of the centroid and the variance and covariance matrix of its variables. If it is assumed that the groups are all described by the same variance and covariance matrix (i.e. all occupy the same shaped area or volume in the feature space) then linear boundaries separating the groups may be relatively easily defined, being located by the likelihood ratio i.e. the ratio of the conditional probability of assigning a pattern x to a class θ_i to that of assigning the pattern to class θ_j:

$$R_{ij} = p(x|\theta_i)/p(x|\theta_j) \qquad (4.4)$$

If we assume that the prior probabilities of assigning a pattern x to any class are equal then we obtain the basic maximum likelihood rule (Steiner & Salerno 1975), that x is assigned to class i, if and only if:

$p(x|\theta_i) > p(x|\theta_j)$ for all $j = 1, 2, \ldots r$, where r is the number of classes (4.5)

If we relax the assumption of equal covariances then quadratic rather than linear discriminants are defined (Steiner 1970). If we have *a priori* knowledge of the likely relative frequency of members in groups, then use can be made of Bayes' decision rule (Brooner *et al.* 1971) which will lead to the location of more efficiently based boundaries when we come to assign the rest of the image to one of the classes. Use of prior probabilities has been successfully proposed as a way of incorporating the effects of relief and other terrain characteristics in improving classification accuracies (Strahler *et al.* 1978, Strahler 1980).

Supervised methods were first applied to and developed for the application of remote sensing to land use and crop surveys (e.g. Marshall *et al.* 1968, Steiner & Maurer 1968, L A R S 1967 and 1968, Bond & Thomas 1973). Subsequently these methods have been applied to more complex problems of terrain evaluation. Examples include work on rangeland classification by Everitt *et al.* (1979); the attempted mapping of geological units in a vegetated area by Watson and Rowan (1971); general purpose mapping of a variety of terrains in arctic and subarctic areas of Canada by Thie *et al.* (1974).

As a more detailed example of the application of supervised methods, work carried out in Yellowstone Park using a twelve-band multispectral scanner, taken from aircraft, is described (Smedes 1971, Smedes *et al.* 1971a). The test area is 31 square kilometres and contains a variety of terrain types and has about 600 m of relief. Training data were collected for the following classes: bedrock exposures, talus, vegetated rock rubble, glacial kame meadow, glacial till meadow, forest, bog, water and shadow areas. For the training areas successful discrimination was achieved using the best four channels with an accuracy of 88% (Table 4.2). Most inaccuracies occurred, not surprisingly, where the terrain units were small. Although the classes were initially chosen because of their visually contrasted appearance rather than because of definite geomorphological/geological needs, these results are clearly encouraging. At least at the level of distinguishing

Table 4.2 Results of terrain mapping using aircraft multispectral scanner data classified by computer-assisted methods, in Yellowstone Park (Smedes 1971, Smedes *et al.* 1972).

Accuracy of recognition in training areas with supervised analysis (%)	
Bedrock	80·0
Talus	81·0
Vegetated rock rubble	91·9
Glacial kame	93·3
Glacial till	74·4
Forest	88·8
Bog	92·7
Water	94·7
Shadows	91·8
Average % accuracy	87·7
Overall accuracy of classification (%)	
Supervised without preprocessing	86
Supervised with preprocessing	88
Non-supervised without preprocessing	80

physiographic units or rock–soil–vegetation association units, useful results apparently can be obtained.

A somewhat different supervised strategy is to derive equations relating the percentage of a particular terrain characteristic as a function of spectral properties and thus to derive data planes of estimates of the characteristic, which can then be subdivided into classes if required subsequently. An example of this approach is by Bentley *et al.* (1976) who estimated the percentage vegetation cover of an area of rangeland vegetation using ratioed Landsat data with an R^2 value of 0·93.

4.10 Unsupervised approach to classification

In unsupervised classification a feature space is used as in the supervised mode, but the ground conditions relating to the image patterns need not be known initially: instead the classes are delimited using the position of image patterns within the feature space. Grouping of patterns within the feature space using some measure of 'nearness' defines clusters, whose position is then used to define class limits. The character of the clusters is determined *a posteriori* by looking at ground properties of samples from each cluster. The resultant classes are not guaranteed to be useful: some of the clusters may be meaningless because they contain too wide a variety of ground conditions.

Literally hundreds of methods of clustering have been developed for a wide variety of purposes apart from pattern recognition in remote sensing (e.g. Sokal & Sneath 1963 in plant taxonomy; Berry 1967 in geographical regionalization). Methods vary according to speed of execution and efficiency with which the clustering takes place. Different criteria of efficiency lead to different approaches. An example of a conceptually simple but not necessarily efficient system is to find which two patterns are closest in the *n*-dimensional feature space, these forming the first cluster; the second two nearest patterns are then grouped and so on; two groups will be linked if any individual members of the groups prove to be the nearest pair at any stage. A more complex scheme would involve replacing the pairs of patterns by group centroids and then using these new points in the search for subsequent nearest pairs; alternatively a pattern might be assigned at any stage to the group from which it has the smallest average distance (e.g. Steiner *et al.* 1969). From the foregoing description it will be clear that the number of groups can vary from one when all image patterns are clustered, to *N*, the number of image patterns. Choice of the number of groups to be defined can be determined using some criterion of separability between the groups, or in many systems it is possible initially to specify the final number of groups or classes.

It has been found useful to combine methods of clustering in order to achieve an efficient solution. Su and Cummings (1972) describe a system consisting of preliminary clustering by a sequential statistical clustering method which requires only one pass through the data to achieve relatively good clustering. A second iterative clustering procedure is then used known as 'Generalized *K*-means clustering' – which clusters very efficiently but depends strongly on an initial good estimate of the location of cluster centres, which is provided by the sequential statistical clustering method.

Like supervised classification, unsupervised methods have commonly been applied to land use and crop surveys (e.g. Steiner *et al.* 1979, Drennan *et al.* 1974). Applications of unsupervised classification include ones for vegetation–ecological studies (Cole *et al.* 1974, Flemming *et al.* 1975, Hoffer & staff 1975); soil studies

(Weismiller, Persinger *et al.* 1977, Seubert *et al.* 1979), for coastal zone land cover types (Paris 1974, Kristof & Weismiller 1977) and for desert resources inventory (Bryant *et al.* 1979).

For the purposes of comparison we can again look at the Yellowstone National Park, USA (Table 4.2) where following the supervised analysis described above (Smedes 1971, Smedes *et al.* 1971a), unsupervised methods were applied both to photographic imagery (Smedes *et al.* 1971b) and to the twelve-channel multispectral data (Smedes *et al.* 1972). Results using the two-stage clustering method of Su and Cummings (1972) described above, give somewhat less satisfactory results than did the supervised approach. Seventeen initial classes were reduced to twelve; these classes corresponded approximately to previously defined categories, with an overall accuracy of 80%. Using colour film as a three-band sensor the test was repeated after digitizing the film data giving rather better results of 85% (Smedes *et al.* 1971b).

The results of an unsupervised classification on an area in southern Italy are presented in Plate 3. The original twenty-five clusters were reduced to ten land cover classes.

4.11 The relative merits of supervised and unsupervised methods

The use of supervised and unsupervised methods represents two very different approaches, one might even say philosophies, towards the problem of automated classification. At first, the supervised approach might appear more valuable since the classes one wishes to discriminate are decided by the human interpreter, who presumably knows the properties he wants mapped. It could be argued with some justification that unsupervised methods can frequently result in the definition of classes of no interest to the interpreter. Furthermore, if unsupervised methods are used then the interpreter will be making less use of ground and ancillary data that are usually available for interpretation. Assuming that no ground data are available, which is unlikely, then unsupervised methods could be regarded as a second best to the supervised approach in at least providing classes of a sort. Clearly the unsupervised approach has special value, when we are unfamiliar with a particular set of imagery or more commonly when we are unfamiliar with its usefulness in a new geographical area. Classes which are defined, should reveal what land cover or terrain types can most successfully be distinguished by using the imagery.

Despite the above, strong arguments have been put forward during recent years stressing the need for applying unsupervised methods, especially within regions with more natural, nonagricultural terrain. Nagy *et al.* (1971) argue against the use of supervised methods even for land use surveys, because of the difficulty of obtaining unique spectral signatures, as a result of differences in, for example, scene illumination, atmospheric conditions, crop maturity, health or variation between species or varieties. They therefore reason that ground data should be collected *after* rather than before analysis particularly in operational systems. Smedes *et al.* (1972) reached similar conclusions in comparing supervised with unsupervised approaches in their analysis of the Yellowstone Park Study area, described above. Additionally the disadvantages of the supervised methods were seen to arise because knowledge of the site was required in order to select training areas and that often the training set might well not be fully representative of the terrain class throughout the test site. Ground data collection for use with the supervised approach therefore tended to be 91 much more time consuming. The resulting maps from unsupervised classification are nearly as accurate as those derived by supervised methods (Table 4.2) but at substantial savings in time and cost.

Work carried out in the rugged San Juan Mountains of Colorado, led to the unsupervised approach being favoured because selection of homogeneous training samples for supervised classification proved extremely difficult as a result of variations in slope and aspect as well as spectral differences in the cover types themselves (Hoffer & staff 1975).

Not surprisingly, given the advantages and disadvantages of the two approaches, intermediate hybrid methods have been suggested. Flemming *et al.* (1975) have proposed both a modified supervised and modified cluster approach. In the former, training areas of known cover type were selected: clustering of the data for each class was then performed which led to the formation of a number of unimodal spectral subclasses for each cover type. This approach helps deal with the problem of highly variable classes alluded to immediately above. In the modified cluster approach, relatively small data blocks from throughout the test area are clustered in a completely unsupervised way. The character of each of the resultant classes is then identified for each of the small data blocks. The results of all training areas are then grouped into clusters according to a separability algorithm to develop a single set of training statistics. Small testing areas are used to assess the success of the classification. Modifications in delimitation of the classes within the feature space are made if necessary and then the entire study area was classified using the maximum likelihood rule. Comparisons between the unsupervised, modified clustered and modified supervised methods using Landsat MSS data were made in the Ludwig Mountain quadrangle using the same 34 test areas (comprising a total of 7844 pixels) (Table 4.3). The modified supervised method yielded the poorest result and the modified cluster method the best. Although the supervised approach was not used, it was believed that it would have yielded almost inevitably poorer results than the modified supervised method.

A further development in the use of hybrid methods is presented by Flemming (1977). Methods are classified according to the procedure used in deriving the training statistics for extrapolation to the whole area from samples using the maximum likelihood rule (Table 4.4). Note that a simple unsupervised clustering of the whole area was not attempted because of the long computer times required. Consideration of accuracy, computer time and time of involvement of the human analyst indicated that the multi-cluster block approach was best.

Choice of the appropriate automated approach will therefore depend on a large number of factors: complicated signatures as a result of complex terrain or a desire to analyze imagery in new situations will all tend to favour an unsupervised approach: in contrast if the terrain has relatively large homogeneous areas or if the class or classes to be distinguished have to be precisely predetermined by the user then a supervised approach will be more appropriate.

Swain and Hausk (1977) have suggested a decision-tree approach in which classification proceeds in successive hierarchical stages using different discriminating properties at each stage. Such an approach is likely to be especially useful when different types of features are used in classification; for example when ancillary data planes are incorporated such as existing geological and soil maps or digital terrain (relief) data. Different stages may involve supervised or unsupervised algorithms within the same classification.

Whichever approach is adopted, the reasons for misclassifications must be sought. These may lie in the limitations of the classification algorithms which were

Table 4.3 Results of classification from the Ludwig Mountain study area (Flemming *et al.* 1975).

Analysis technique	% correctly classified
Modified supervised	70·0
10 class non-supervised	76·6
20 class non-supervised	78·5
Modified cluster	84·7

Four cover types, including agriculture, water, deciduous forest and coniferous forest.

Table 4.4 Summary of results of alternative computer-aided analysis procedures of forest lands in the San Juan Mountains, Colorado (from Flemming 1977).

Sample used in production of training statistics	Type of clustering		
	No clustering	Multiple clustering	Single clustering
Statistical sample			Unsupervised AT 26 hours CT 48 minutes Acc. 76% TSP Clustering of whole sample
Heterogeneous sample blocks containing several cover types		Multi-cluster blocks AT 12 hours CT 22 minutes Acc. 78·8% TSP Selection of representative sample blocks. Separate clustering of each block.	Mono-cluster blocks AT 10·5 hours CT 25 minutes Acc. 73·1% TSP Selection of representative sample blocks. Clustering of sample blocks combined.
Homogeneous training fields (sites) containing single cover types	Supervised AT 53 hours CT 25 minutes Acc. 64·7% TSP Training sites for each class selected. Sites of each class combined to derive mean statistics.	Multi-cluster training fields AT 50 hours CT 19 minutes Acc. 69·7% TSP Training sites for each class combined. Clustering of each set of training sites	Mono-cluster training fields AT 45·5 hours CT 22 minutes Acc. 69·4% TSP Training sites for all classes combined, and clustered.

AT: analyst time   CT: computer time   Acc.: accuracy
TSP: Training set procedures (method for creating final classes for production of the training statistics which are used for extrapolation to the whole target area).

applied or in the features which were selected: in such cases improvements can probably be made with the existing data sets. Misclassifications may have arisen from failure adequately to pre-process the data or to select ground sites for testing and training with sufficient rigour. It may be that the analyst finds the explanation is in the inherent spectral, spatial and radiometric resolving capabilities of the sensors in relation to a particular type of terrain. A thorough assessment of performance is clearly important for the analyst to plan a strategy to obtain improved results.

4.12 Recognition and classification of shapes within images

An element missing from the classification procedures described above is the importance of recognizing parts of the images as distinct objects (Prewitt 1970) with specific geometrical properties. Tunstall (1975) has suggested that this procedure is an absolutely fundamental precursor of all human pattern recognition. For example, a river is more likely to be recognized by its meander shapes than the spectral or textural properties produced by the water surface. Considerable progress has been made in the character recognition field such that operational systems for recognizing handwriting are now in existence. This approach has only been used to a comparatively minor extent in the interpretation of remotely sensed imagery. Three types of shapes may be recognized, namely point, line and areal forms. Extrapolation to three-dimensional and higher-dimensional volumes is conceivable if we considered object shape in different spectral bands and different times, though recognition of identifiable homogeneous objects would be an increasingly severe problem.

Distinguishing point objects is feasible by automated means (Gonzalez & Wintz 1977). Operational systems include that devised in the Quantimet range of instruments (Wignall 1975).

Although a large number of statistical methods exists for describing point patterns (Sayn-Wittgenstein 1970, Dacey 1963), the applicability of methods is likely to be restricted to specific scales of images and to the study of certain vegetation patterns and certain types of terrain which possess a substantial number of quasi 'point'-like objects as, for example, in some types of limestone terrain and some periglacial landscapes.

Recognition of linear features, would seem to be potentially more fruitful because of the dominance of fluvial erosion in many of the world's landscapes. Applications in recognizing man-made features such as roads suggest these methods could be useful. As an example we refer to the work of VanderBrug (1976) who has carried out a comparison of the performance of linear, semi-linear and non-linear filters (Fig. 4.21) for detecting linear features on imagery both from the Skylab 190B camera and Landsat 1 multispectral scanner. He found that all proved encouraging and that a nonlinear detector was best. The nonlinear detector produced thinner lines than the semilinear, which also responded more strongly to background noise. Subsequently Gurney (1980) has shown the advantage of combining elements of the nonlinear and semilinear detectors.

Areal object-seeking algorithms usually exploit the internal homogeneity of objects and their contrast with their surroundings, though if textural properties are used then homogeneity of tones will not be the principal homogeneity criterion, which will instead be some sort of characteristic heterogeneity (Stanley 1977). A variety of methods for recognition of objects are available, either by progressive subdivisions of images, namely disjunctive methods (Robertson 1973), or by

Given a 3 × 3 array:

$$A_1 \quad B_1 \quad C_1$$

$$A_2 \quad B_2 \quad C_2$$

$$A_3 \quad B_3 \quad C_3$$

For a vertical bright line:

(a)   linear detector: $\Sigma B_i > \dfrac{\Sigma C_i + \Sigma A_i}{2}$ (Prewitt 1970)

(b)   non-linear detector: $\left\{ B_2 > A_2 \text{ and } B_2 > C_2 \right\}$ (Rosenfeld & Thurston 1971)

$\left\{ B_1 > A_1 \text{ and } B_1 > C_1 \right\}$

$\left\{ B_3 > A_3 \text{ and } B_3 > C_3 \right\}$

(c)   semi-linear detector: $\Sigma B_i > \Sigma A_i$ (VanderBrug 1976)

$\Sigma B_i > \Sigma C_i$

(d)   combined non-linear and semi-linear detector: $\left\{ B_2 > A_2 \text{ and } B_2 > C_2 \right\}$ (Gurney 1980)

$\left\{ \Sigma B_i > \Sigma A_i \text{ and } \Sigma B_i > \Sigma C_i \right\}$

Figure 4.21 Types of detectors of lines in digital images.

progressive amalgamation, namely conjunctive methods (Kettig & Landgrebe 1976). Further discussion of this topic is presented in the next chapter on regionalization.

4.13 Use of contextual information

Contextual information plays a central role in nearly all human interpretation but at present has been used relatively rarely in automated systems. Contextual information comprises information outside of an image element which aids its identification. We distinguish it from texture which simply describes the spatial variations of picture point values but does not involve previous recognition or classification of the surroundings.

Several types of contextual information can readily be identified. Firstly we should distinguish between what Welch and Salter (1971) call *presence indicators* and *appearance modifiers*. The latter includes such features as snow cover, atmospheric haze and, in some circumstances, shadows: Welch and Salter also include, to our minds inappropriately, such factors as dynamic range of film. However, *presence indicators* are of more general interest and four main types may be recognized.

(a) *Extra-image context indicators* include knowledge of properties such as general geographical location, time of acquisition and scale of imagery, all of which are

95

of enormous help in human interpretation since these will allow the experienced interpreter immediately to be aware of the likely range of both the surface features and the subsurface features of which they are an expression.

(b) *Higher-order context indicators* occur where recognition of an object within a scene aids identification of its component parts. For example, a human interpreter, having recognized a flood plain, may then be able to identify many of its more poorly displayed components such as vegetated abandoned channels and palaeo-point bars.

(c) *Equal-order context indicators* are where recognition of one part of a scene helps the recognition of similar objects which are more poorly displayed elsewhere. For example, a human interpreter can often recognize a river or a road as a continuous object by starting with easily identified parts and extending them into parts of the scene where they are less clearly depicted. In automated schemes this principle can be used successfully to remove inliers from otherwise homogeneous areas, on the supposition that a pixel is more likely to belong to the class of pixels surrounding it. This is likely to work especially well where the ground features are large relative to the picture point size and thus large homogeneous objects are common in the scene.

(d) *Lower-order context indicators* are where the recognition of the components in a scene helps recognition of the object of which they are a part. Thus using the example of the flood plain mentioned above, it may well be first recognized by well displayed components such as actively meandering channels and recently abandoned ox-bow lakes.

Some attempts have been made to use context in an automated approach. Fu (1976) has applied a structural or syntatic approach to pattern recognition which takes into account the hierarchical character of contextual information (Section 5.3.2).

Welch and Salter (1971) derived a method for analysing context using the Bayes decision rule, the simple non-contextual form of which involves minimization of:

$$\sum_{\theta=1}^{r} L(\theta,a)\, p(x|\theta)\, G(\theta) \qquad (4.6)$$

where $L(\theta,a)$ is the loss incurred in assigning a given picture cell to an *a priori* class θ using decision rule a; $p(x|\theta)$ is the probability density function of the pattern x belonging to class θ and $G(\theta)$ is the prior probability of occurrence of class θ, and r is the number of classes. The above minimization will lead to a minimization of the average loss not using context.

This expression can be modified to take into account other picture cells to aid identification of any one cell giving:

$$\sum_{\theta_k=1}^{r} L(\theta_k,a_k)\, p(x_n|\theta_k) G(\theta_k) \qquad (4.7)$$

where the only difference is that all of the recognition vectors from the frame are used, while the simple decision rule uses only the cell k.

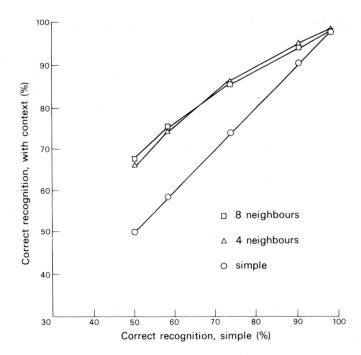

Figure 4.22 Plot of percent correct recognition with context versus percent correct recognition without context. The results are shown for five different levels of simple recognition accuracy for both four and eight neighbour rules, using simulated date with twenty-two dimensional vectors to give the different levels. When simple recognition is better than 80% correct, the error rate is reduced by one half and when the simple recognition rate is lower than this value, the improvement in accuracy is at least 10% (from Welch & Salter 1971).

For most pictures the above expression is impractical because of the large number of other picture cells, so that Welch and Salter (1971, p. 26) only considered the four immediately adjacent cells and that no appearance modification was present. This yields the following expression which needs to be minimized:

$$\sum_{\theta_k=1}^{r} L(\theta_k, a) p(x_k | \theta_k) G(\theta_k) \prod_{i=1}^{4} \sum p(x_{ki} | \theta_{ki}) . p(\theta_{ki} | \theta_k) \qquad (4.8)$$

whose left-hand part is identical to the simple rule. The product term contains the context contribution from the four adjacent cells. Results were encouraging as shown in Figure 4.22, but are difficult to evaluate properly since although real imagery was used, the actual measurements or recognition data were simulated. More recent work by Swain *et al.* (1979) also suggests context classifiers can improve classification accuracies.

4.14 Monitoring change

Our previous discussion has concentrated, implicitly at least, on base-line inventories of terrain characteristics, but one of the principal potential benefits of remote sensing is its monitoring capability. In realizing this potential, we are often hindered by the fact that there are frequently many differences between images other than those indicative of the terrain characteristics we wish to monitor. For example, differences due to contrasts in Sun elevation and azimuth, and changes in atmospheric conditions, will need to be minimized. Moreover, a further difficulty is the confusing effect of changes in terrain characteristics occurring concurrently with those being monitored. Such changes may operate over very different time scales, from diurnal, through seasonal, annual, and much longer periods. To help overcome some of these problems, images with similar calendar dates acquired at the same time of day are normally used.

In general, successful monitoring with remote sensing can only take place if we consider the spatial and temporal resolutions of the sensing systems in relation to the spatio-temporal characteristics of the phenomenon we are monitoring, as shown diagrammatically in Figure 4.23 (Townshend 1977).

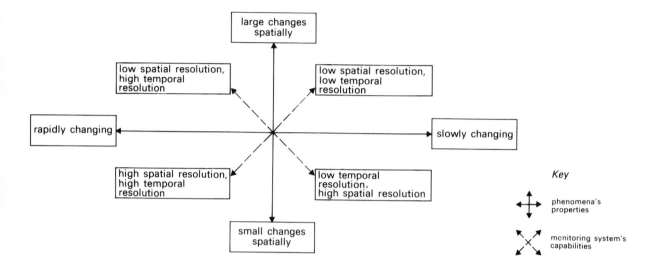

Figure 4.23 Relationship between the capabilities of remote sensing monitoring systems and the spatio-temporal properties of the phenomena being monitored (Townshend 1977).

Robinson (1979) has produced a useful review of the approaches used for detecting change by means of remote sensing. First, there are differencing techniques which involve registration of images from two dates, subtraction of the pixel values of one from the other and then analysis and classification of the resultant differenced image. The images used for differencing may be the original images or consist of ratios of channels (Section 4.2.2) or even principal components (Angelici *et al.* 1977). Alternatively, images from the two dates can be expressed as a ratio and the size of the deviations from unity used to classify the degree and type of change. A related method is to derive the regression line between the pixel values of the second date against the first and classify according to the size of the deviations from the regression line. The latter method has the particular advantage of directly compensating for overall differences in brightness, by the size of the intercept. A contrasting approach is to register two sets of multiband images and then carry out an unsupervised classification on the resultant single set of images, on the basis that areas which have changed will form distinctive clusters in the feature space (Section 4.10). This method was included in one of the few comparative analyses of change detection methods by Weismiller, Kristof *et al.* (1977) in a study of coastal change in southern Texas. However, on the basis of a somewhat subjective appraisal they concluded that classifying each set of images separately and then differencing the classified maps was preferable. However, Robinson (1979) rightly points out that methods involving classification of all the pixels in an image rather than restricting attention to those displaying notable changes will be computationally much more demanding.

98

4.15 Man–machine interactive systems

The 1970s saw the introduction of computer-based systems specifically designed for image processing (Lillestrand & Hoyt 1974). These do not necessarily involve any new algorithms, but are an improvement on batch-mode main frame computers since they allow the interpreter to see the results of his enhancements and classifications and to modify his decisions rapidly. In this way the value of the

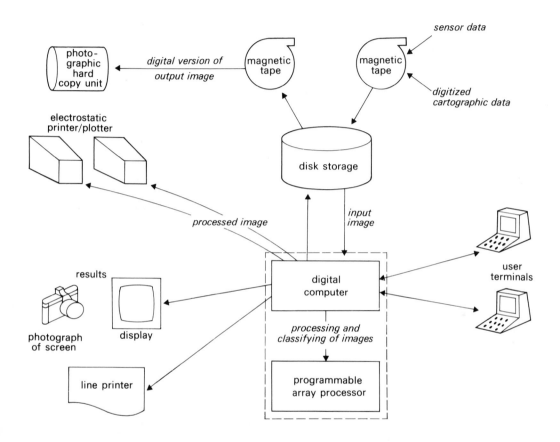

imagery can be rapidly exploited, and image manipulations carried out with speeds compatible with a human analyst exerting his high level interpretation skills. Thus the benefits of the computational rapidity of the computer are combined with the intellectual skills and background or contextual knowledge of the human interpreter.

Figure 4.24 Components of an interactive image processing system. The ESL Interactive Digital Image Manipulation System (IDIMS) (from ESL 1976).

The layout of a typical system is shown in Figure 4.24. Input is from sensor data and digitized cartographic data. The computer is operated by the analyst directly from user terminals and he can view the results of his work directly on a colour display screen. Hard copy output can be obtained crudely by photography from the screen, or from an electrostatic plotter giving a grey-scale image or when final quality products are required a photographic hard copy unit is used.

In the particular example illustrated, computational speeds are improved by use of a programmable array processor as in the IDIMS system (ESL 1976). Alternatively, hard-wire devices have been incorporated, designed to execute

99

specific algorithms as in the G E-100 (General Electric Company 1976). Although very fast the latter have the limitation of being unchangeable without physical reconstruction, so that improvements in analytical techniques are difficult to include (Landgrebe 1976). If a computer is specially dedicated to image processing and classification and software specifically written for it, then a reasonable degree of interaction is possible even when using a mini-computer (Mulder & Donker 1977).

Examples of the use of these devices appear in several accounts (Kriegler *et al.* 1975, Economy *et al.* 1974, Lefebvre 1975, Townshend & Justice 1980).

4.16 Conclusion

Image processing and classification by machine-assisted methods can greatly enhance the value of remotely sensed data. Their use however makes many demands. Firstly, they will always add to the cost of interpretation. Special purpose image processing equipment in particular is often expensive either to purchase or rent. The interpreter will need to evaluate whether the benefits are worthwhile for a particular task. Such evaluations will themselves involve investment. Secondly, training of analysts will be necessary. Such training falls into two separate categories. The theory of image processing and classification has to be learnt, including such concepts as feature space, point processing, supervised and unsupervised classification and so on. In contrast there is the training needed to implement the theory. This may be in advanced photographic methods, in programming computers or in operating digital image processing systems by means of special high-level programming languages. Both types of training are essential if useful results are to be obtained. The magnitude of the task will be daunting for many concerned with terrain resources evaluation. Nevertheless, there is little to be gained and much to be lost in teaching operators to implement advanced pattern enhancement and classification algorithms and digital systems if they do not have an understanding of the underlying theoretical principles.

References

Abrosimov, I. K. and Y. M. Kleiner 1973. The use of landscape-indicator methods in hydrogeological investigations. In *Landscape indicators*, A. G. Chikishev (ed.), 34–8. New York: Consultants Bureau.

Aggarwal, J. K., R. O. Duda and A. Rosenfeld 1977. *Computer methods in image analysis*. New York: I E E E Press.

Algazi, V. D. 1973. Multispectral combination and display of ERTS-1 data. *3rd Earth Resources Technology Satellite-1 Symp.*, NASA SP-351, 1709–17. Washington, D C: NASA.

American Society of Photogrammetry 1960. *Manual of photographic interpretation*. Washington, D C: American Society of Photogrammetry.

American Society of Photogrammetry 1968. *Manual of colour aerial photography*. Virginia: American Society of Photogrammetry.

Angelici, G. L., N. A. Bryant and S. Z. Friedman 1977. Techniques for land use change detection using Landsat imagery. *Proc. Am. Soc. Photogramm.* 217–28.

Arsenault, H. H., M. K. Seguin, N. Brouseau and G. April 1974. Le traitement optique des images ERTS. *Proc 2nd Canad. Symp. on Remote Sensing, Guelph, Ontario, Canada*, **II**, 488–94.

Bandat, H. F. von 1962. *Aerogeology*. Houston, Texas: Gulf Publishing Co.

Barnett, M. E. and R. P. Harnett 1977. Optical processing as an aid in analysing remote sensing imagery. In *Environmental remote sensing* **2**, E. C. Barrett and L. F. Curtis (eds), 124–42. London: Edward Arnold.

Barnett, M. E. and T. H. Williams 1979. Video-rate image processing. In *Electronic imaging*, T. P. McClean and P. Schager (eds). London: Academic Press.

Batson, R. M., K. Edwards and E. M. Eliason 1976. Stereo and Landsat Pictures. *Photogramm. Engng* **42**, 1279–84.

Bauer, A., A. Fontanel and G. Grau 1967. The application of optical filtering in coherent light to the study of aerial photographs of Greenland glaciers. *J. Glaciol.* **6**, 781–93.

Beaumont, T. E. 1977. *Techniques for the interpretation of remote sensing imagery for highway engineering purposes*. TRRL Rep. LR753, Transport and Road Research Laboratory, Crowthorne: Department of the Environment.

Bentley, R. G., B. C. Salmon-Dreyler, W. J. Booner and R. K. Vincent 1976. *A Landsat study of ephemeral and perennial rangeland vegetation and soils*. US Department of the Interior, Bureau of Land Management.

Bernstein, R. 1978. *Digital image processing for remote sensing*. New York: IEEE Press.

Berry, B. J. L. 1967. Grouping and regionalisation. In *Quantitative geography*, part 1, W. L. Garrison and D. Marble (eds). North West University Studies in Geography, no. 13.

Bodechtel, J. and G. Kritikos 1971. Quantitative image enhancement of photographic and non-photographic data for earth resources. *Proc. 7th Int. Symp. on Remote Sensing of Environment, Ann Arbor, Michigan*, 469–86.

Bond, A. D. and D. T. Thomas 1973. Automatic information extraction for land use and agricultural purposes. *Proc. Symp. on Management and Utilization of Remote Sensing Data*, 257–67. American Society of Photogrammetry.

Brooks, R. R. 1972. *Geobotany and biogeochemistry in mineral exploration*. New York: Harper & Row.

Brooner, W. G., R. M. Haralick and T. Dinstein 1971. Spectral parameters affecting automated image interpretation using Bayesian probability techniques. *Proc. 7th Int. Symp. on Remote Sensing of Environment, Ann Arbor, Michigan*, **III**, 1929–49.

Bryant, N. A., R. G. McCleod, A. L. Zobrist and H. B. Johnson 1979. California desert resource inventory using multispectral classification of digitally mosaicked Landsat frames. *Proc. 1979 Machine Processing of Remotely Sensed Data Symp., Purdue University, Indiana*, 69–78.

Burns, K. L., J. Shepherd and N. Berman 1977. Reproducibility of geological lineaments and other discrete features interpreted from imagery: measurement by a coefficient of association. *Remote Sensing of Environment* **5**, 267–301.

Carter, L. D. and R. O. Stone 1974. Interpretation of orbital photographs. *Photogramm. Engng* **40**, 193–202.

Coiner, J. C. 1972. *SLAR image interpretation keys for geographic analysis*. CRES Tech. Rep. 177–19, University of Kansas, Lawrence, Kansas.

Cole, M. M., E. S. Owen-Jones, N. D. E. Custance, and T. E. Beaumont 1974. Recognition and interpretation of spectral signatures of vegetation and associated environmental parameters from aircraft and satellite imagery in W. Queensland, Australia. In *European Earth Resources Satellite Experiments, Frascati, Italy*, B. T. Battrick and N. T. Duc (eds), ESRO SP-100, 243–87.

Dacey, M. F. 1963. Order neighbor statistics for a class of random patterns in multidimensional space. *Ann. Assn Am. Geogrs* **53**, 505–15.

Dalrymple, J. R., R. J. Blong and A. J. Conacher 1968. A hypothetical nine unit land surface model. *Z. Geomorph.* **12**, 60–76.

Dethier, B. E. 1974. *Phenology satellite experiment*. Final proj. rep., NASA, contract NAS 5-21781.

Donker, N. H. W. and A. M. J. Meijerink 1977. Digital processing Landsat imagery to produce a maximum impression of terrain ruggedness. *ITC J.* **4**, 683–704.

Donker, N. H. W. and N. J. Mulder 1976. Analysis of MSS digital imagery with the aid of principal component transform, ISP Commission VII. *13th Cong. Int. Soc. Photogramm.*

101

Drennan, D. S. H., C. J. Bray, I. R. Galloway, J. R. Hardy, C. O. Justice, E. S. Owen-Jones, R. A. G. Savigear and J. R. G. Townshend 1974. The interpretation of false-colour infrared and true colour photography of part of Argentina obtained by Skylark Earth Resources rockets. *Proc. 9th Int. Symp. on Remote Sensing of Environment, Ann Arbor, Michigan*, 1475–96.

Duda, R. O. and P. E. Hart 1973. *Pattern classification and scene analysis.* New York: John Wiley.

Eardley, A. J. 1941. *Aerial photographs: their use and interpretation.* New York: Harper & Row.

Economy, R., D. Goodenough, R. Ryerson and R. Towles 1974. Classification accuracy of the Image 100. *Proc. 2nd Canad. Symp. on Remote Sensing, Guelph, Ontario, Canada*, 278–87.

ESL Inc. 1976. *IDIMS user's guide, technical memorandum.* ESL-TM 705, Sunnyvale, California.

Estes, J. E. and D. Simonett 1975. Fundamentals of image interpretation. In *Manual of remote sensing*, R. G. Reeves (ed.), 869–1076. Virginia: American Society of Photogrammetry.

Evans, R., J. Head and M. Dirkzwager 1976. Air photo-tones and soil properties: implications for interpreting satellite imagery. *Remote Sensing of Environment* **4**, 265–80.

Everitt, J. H., A. J. Richardson, A. Gerbermann, C. L. Wiegand and M. A. Alaniz 1979. Landsat 2 data for inventorying rangelands in South Texas. *Proc. 1979 on Machine Processing of Remotely Sensed Data Symp. Purdue University, Indiana*, 132–40.

Flemming, M. D. 1977. *Computer-aided analysis techniques for an operational system to map forest lands utilizing Landsat MSS data.* LARS Tech. Rep. 112277. West Lafayette, Indiana: Laboratory for Applications of Remote Sensing, Purdue University.

Flemming, M. D., J. S. Berkebile and R. M. Hoffer 1975. Computer-aided analysis of Landsat-1 MSS data: a comparison of 3 approaches including a 'modified clustering' approach. *Proc. 1975 Machine Processing of Remotely Sensed Data Symp.*, 1B-54 to 1B-62.

Fraser, R. S. 1974. Computed atmospheric corrections for satellite data. *Proc. Soc. Photo-Optical Instrumentation Engrs* **51**, 64–72.

Frazee, C. J., R. L. Carey and F. C. Westin 1972a. Utilizing remote sensing data for land use decisions for Indian Lands in South Dakota. *Proc. 8th Int. Symp. on Remote Sensing of Environment, Ann Arbor, Michigan*, **1**, 375–91.

Frazee, C. J., V. I. Myers and F. C. Westin 1972b. Density slicing techniques for soil survey. *Soil Sci. Soc. Am. Proc.* **36**, 693–5.

Fridland, V. M. 1976. *Pattern of the soil cover.* Academy of Sciences of USSR, transl N. Kaner, Jerusalem: Keter.

Fu, K.-S. 1976. Pattern recognition in remote sensing on the Earth's resources. *IEEE Trans. Geosci. Electron.* **GE-14**, 10–18.

General Electric Company 1976. *Image 100 interactive multispectral image analysis system: system description.* Daytona Beach, Florida.

Goetz, A. F. H., F. C. Billingsley, A. R. Gillespie, M. J. Abrams, R. L. Squires, E. M. Shoemaker, I. Lucchilta and P. Elston 1975. *Application of ERTS images and image processing to regional geologic problems and geologic mapping in Northern Arizona.* Jet Propulsion Laboratory, Pasadena, California, Tech. Rep. 32-1597.

Gonzalez, R. C. and M. G. Thomason 1978. *Syntactic pattern recognition.* Reading, Mass.: Addison-Wesley.

Gonzalez, R. C. and P. Wintz 1977. *Digital image processing.* Reading, Mass.: Addison-Wesley.

Goodman, J. W. 1968. *Introduction to Fourier optics.* San Francisco: McGraw-Hill.

Gramenopoulos, N. 1974. Automated thematic mapping and change detection of ERTS-1 images. *3rd Earth Resources Technology Satellite-1 Symp.*, **1B**, NASA SP-351, Washington, DC, 1845–75.

Grasselli, A. 1969. *Automatic interpretation and classification of images.* London: Academic Press.

Gurney, C. M. 1980. Threshold selection for line detection algorithms. *IEEE Trans. Geosci.*

Haralick, R. M. 1973. Glossary and index to remotely sensed image pattern concepts. *Pattern Recognition* **5**, 391–403.

Haralick, R. M. and R. Bosley 1974. Kansas environmental and resource study: *A Great Plains model. Spectral and textural processing of ERTS imagery*. NASA Contract no. NAS5-21822. Lawrence, Kansas: University of Kansas.

Haralick, R. M. and K. S. Shanmugam 1974. Combined spectral and spatial processing of ERTS imagery data. *Remote Sensing of Environment* **3**, 3–13.

Haralick, R. M., K. S. Shanmugam and I. Dinstein 1973. Textural features for image classification. *IEEE Trans. Syst., Man, Cybern.* **SMC-3**, 610–21.

Harnett, P. R., G. D. Mountain and M. E. Barnett 1978. Spacial filtering applied to remote sensing images, *Optica Acta* **25**, 801–809.

Hoffer, R. M. and staff 1975. *Computer-aided analysis of Skylab multispectral scanner data in mountainous terrain for land use, forestry, water resource, and geologic applications*. LARS Information Note 121275. Laboratory for Applications of Remote Sensing, Purdue University, West Lafayette, Indiana.

Holben, B. N. and C. O. Justice 1980. *An examination of spectral band ratioing to reduce the topographic effect on remotely sensed data*. NASA/GSFC Tech. Mem. 80640.

Hornung, R. J. and J. A. Smith 1973. Application of Fourier analysis to multispectral spatial recognition. *Symp. Proc. Management and Utilization of Remote Sensing Data, Sioux Falls, South Dakota*, 268–83.

Humiston, H. A. and G. E. Tisdale 1973. A peripheral change detection process. *Symp. Proc. Management and Utilization of Remote Sensing Data, Sioux Falls, South Dakota*, 413–26.

Hunt, B. R. 1975. Digital image processing. *Proc. IEEE* **63**, 693–708.

Jensen, N. 1973. High-speed image analysis techniques. *Photogramm. Engng* **39**, 1321–8.

Justice, C. O. 1978. An examination of the relationships between selected ground properties and Landsat MSS data in an area of complex terrain in southern Italy. *Proc. Amer. Soc. Photogramm., Fall Meeting, Albuquerque, New Mexico*, 407–21.

Justice, C. O., D. F. Williams, J. R. G. Townshend and R. A. G. Savigear 1976. The evaluation of space imagery for obtaining environmental information in Basilicata Province, S. Italy. In *Land use studies by remote sensing*, W. G. Collins and J. L. van Genderen (eds), 1–39. Remote Sensing Society.

Kettig, R. L. and D. A. Landgrebe 1976. Classification of multispectral image data by extraction and classification of homogeneous objects. *IEEE Trans. Geosci, Electron.* **GE-14**, 19–26.

Koopmans, B. N. 1974. Should stereo SLAR imagery be preferred to single strip imagery for thematic mapping? *Proc. Symp. on Remote Sensing and Photo-interpretation* **II**, 841–61. Alberta, Canada: International Society for Photogrammetry.

Kriegler, F. J., M. F. Gordon, R. H. McLaughlin and R. E. Marshall 1975. The Midas Processor. *Proc. 10th Int. Symp. on Remote Sensing of Environment, Ann Arbor, Michigan*, 757–67.

Kristof, S and R. A. Weismiller 1977. *Computer-aided analysis of Landsat data for surveying Texas coastal zone environments*. LARS Tech. Rep. 090677, Laboratory for Applications of Remote Sensing, West Lafayette, Indiana.

Krumbein, W. C. and F. A. Graybill 1965. *An introduction to statistical models in geology*. New York: McGraw-Hill.

Landgrebe, D. 1976. Computer-based remote sensing technology: a look to the future. *Remote Sensing of Environment* **5**, 229–46.

LARS (Laboratory for Agricultural Remote Sensing) 1967. *Remote multispectral sensing in agriculture*. Report of LARS, 2, Purdue University, Agric. Exp. St. Res. Bull. 832, West Lafayette, Indiana.

LARS 1968. *Remote multispectral sensing in agriculture*. Report of LARS, 3, Purdue University, Agric. Exp. St. Res. Bull. 844, West Lafayette, Indiana.

Lee, Y. J., E. T. Oswald and J. W. E. Harris 1974. A preliminary evaluation of ERTS imagery for forest land management in British Columbia. *Proc. 2nd Canad. Symp. on Remote Sensing, Guelph, Ontario*, **I**, 88–101.

Lefebvre, R. H. 1975. Mapping in the craters of the Moon volcanic field, Idaho with Landsat (ERTS) imagery. *Proc. 10th Int. Symp. on Remote Sensing of Environment, Ann Arbor, Michigan*, 951–63.

Lillestrand, R. L. and R. R. Hoyt 1974. The design of advanced digital image processing systems. *Photogramm. Engng* **40**, 1201–18.

Leuder, D. R. 1959. *Aerial photographic interpretation*. New York: McGraw-Hill.

Malan, O. G. 1976. *How to use transparent diazo colour film for interpretation of Landsat images*. COSPAR Technique Manual Series, Manual 6.

Malila, W. A., R. F. Nalepka and J. E. Sarno 1975. *Final report: image enhancement and advanced information extraction techniques for ERTS-1 data*. Environmental Research Institute of Michigan, Ann Arbor.

Marshall, R. E., N. Thomson, F. Thomson and E. Kriegler 1969. Use of multispectral recognition techniques for conducting rapid, wide-area wheat surveys. *Proc. 6th Int. Symp. on Remote Sensing of Environment, Ann Arbor, Michigan*, **I**, 3–32.

Mather, P. M. 1976. *Computational methods of multivariate analysis in physical geography*. London: John Wiley.

Meijerink, A. H. J. and N. H. W. Donker 1978. The ITC approach to digital processing applied to land use mapping in the Himalayas and Central Java. In *Remote sensing applications in developing countries*, W. G. Collins and J. L. van Genderen (eds), 75–83. Remote Sensing Society.

Merifield, P. M. and D. L. Lamar 1975. Enhancement of geologic features near Mojave, California by spectral band ratioing of Landsat and MSS data. *Proc. 10th Int. Symp. on Remote Sensing of Environment, Ann Arbor, Michigan*, 1067–75.

Meyer, W. and R. I. Welch 1975. Water resources assessment. In *Manual of remote sensing*, R. G. Reeves (ed.), 1479–551. Virginia: American Society of Photogrammetry.

Miller, V. C. and C. F. Miller 1961. *Photogeology*. New York: McGraw-Hill.

Miller, V. C. and S. A. Schumm 1964. *Proc. Conf. on Aerial Surveys and Integrated Studies, Toulouse*, 41–79. Unesco.

Morgan, O. E. 1977. Data reduction and information extraction for remotely sensed images. *J. Br. Interplanetary Soc.* **30**, 193–200.

Mulder, N. J. and N. H. W. Donker 1977. Poor man's image processing – a stimulus in thinking. *Int. Symp. on Image Processing*. Graz, Austria: International Society of Photogrammetry, Commission III.

Nagy, G., G. Shelton and J. Toloba 1971. Procedural questions in signature analysis. *Proc. 7th Int. Symp. on Remote Sensing of Environment, Ann Arbor, Michigan*, 1387–401.

NAS (National Academy of Sciences) 1977. *Resource sensing from space: prospects for developing countries*. Washington, DC: National Research Council.

Newton, A. P. 1973. Pseudo stereoscopy with radar imagery. *Photogramm. Engng* **39**, 1055–8.

Nielsen, U. 1974. Interpretation of ERTS-1 imagery aided by photographic enhancement. *3rd Earth Resources Technol. Satellite-1 Symp.* **1B**, SP-351, Washington, DC, 1733–86.

Osipova, O. A. 1973. The significance of altitudinal zones in rock-indicator investigations in mountains. In *Landscape indicators*, A. G. Chikishev (ed.), 29–33. New York: Consultants Bureau.

Owen-Jones, E. S. 1977. Densitometer methods of processing remote sensing data, with special reference to crop-type and terrain studies. In *Environmental remote sensing 2*, E. C. Barrett and L. F. Curtis (eds), 101–24. London: Edward Arnold.

Paris, J. F. 1974. Coastal zone mapping from ERTS-1 data using computer-aided techniques. *Proc. 2nd Canad. Symp. on Remote Sensing, Guelph, Ontario, Canada*, 516–28.

Parsons, C. L. and G. M. Jurica 1975. *Correction of Earth resources technology satellite multispectral scanner data for the effect of the atmosphere*. LARS Inf. Note 061875. Laboratory for Applications of Remote Sensing, Purdue University, West Lafayette, Indiana.

Perrin, R. M. S. and C. W. Mitchell 1969. *An appraisal of physiographic units for predicting site conditions in arid areas*, 2 vols. MEXE rep. no. 1111. Christchurch, Hampshire: Military Engineering Experimental Establishment.

Peterson, R. 1979. Oil and gas exploration by pattern recognition of lineament assemblages associated with bends in wrench faults. *Proc. 13th Int. Symp. on Remote Sensing of Environment, Ann Arbor, Michigan*, 993–1014.

Peterson, R. M., C. R. Cochrane, S. A. Morain and D. S. Simonett 1969. A multisensor study of plant communities at Horsefly Mountain, Oregon. In *Remote sensing in ecology*, P. Johnson (ed.), 63–93. Georgia: Georgia University Press.

Pincus, H. J. 1969. The analysis of remote sensing displays by optical diffraction. *Proc. 6th Int. Symp. on Remote Sensing of Environment, Ann Arbor, Michigan*, 261–74.

Prewitt, J. M. S. 1970. Object enhancement and extraction. In *Picture processing and psychopictorics*, B. S. Lipkin and A. Rosenfeld (eds), 75–149, New York: Academic Press.

Poulton, C. E. 1975. Range resources: inventory, evaluation and monitoring. In *Manual of remote sensing*, R. G. Reeves (ed.), 1427–78. Virginia: American Society of Photogrammetry.

Poulton, C. E. 1973. The advantages of side-lap stereo interpretation of ERTS-1 imagery in northern latitudes. *ERTS-1 Symp. Proc., Goddard Space Flight Center, NASA*, 157–61.

Ranz, E. and S. Schneider 1971. Progress in the application of Agfa contour equidensity film for geo-scientific photointerpretation. *Proc. 7th Int. Symp. on Remote Sensing of Environment, Ann Arbor, Michigan*, 779–90.

Rohde, W. F. and C. E. Olson 1970. Detecting tree moisture stress. *Photogramm. Engng* **36**, 561–6.

Robertson, T. V. 1973. Extraction and classification of objects in multispectral images. *Proc. Conf. on Machine Processing of Remotely Sensed Data, Purdue University, Indiana*, Sec. 3B, 27–34.

Robinson, J. W. 1979. *A critical review of the change detection and urban classification literature.* CSC/TM-79/6235, Goddard Space Flight Center. Maryland: Computer Sciences Corporation.

Rosenfeld, A. 1969. *Picture processing by computer.* New York: Academic Press.

Rosenfeld, A. 1976. *Digital picture analysis.* Berlin: Springer-Verlag.

Rosenfeld, A. and M. Thurston 1971. Edge and curve detection for visual scene analysis. *IEEE Trans. Computing* **C-205**, 562–9.

Sabins, F. F. 1978. *Remote sensing principles and interpretation.* San Francisco: W. H. Freeman.

Sayn-Wittgenstein, L. 1970. Patterns of spatial variation in forests and other natural populations. *Pattern Recognition* **2**, 245–53.

Schlosser, M. S. 1974. Television scanning densitometer. *Photogramm. Engng* **40**, 199–202.

Seubert, C. E., M. F. Baumgardner and R. A. Weismiller 1979. Mapping and estimating areal extent of severely eroded soils of selected sites in northern Indiana. *Proc. 1979 Machine Processing of Remotely Sensed Data Symp., Purdue University, Indiana*, 234–9.

Schlien, S. and S. Smith 1975. A rapid method to generate spectral thematic classification of Landsat imagery. *Remote Sensing of Environment* **4**, 67–77.

Short, N. M., P. D. Lowman Jr, S. C. Freden and W. A. Finch Jr 1976. *Mission to Earth: Landsat views the world.* Washington, DC: NASA.

Shrumpf, B. J. 1975. *Multiseasonal-multispectral remote sensing of phenological change for natural vegetation inventory.* Unpub. PhD Thesis, Oregon State University, Xerox University Microfilm 75-26, 006.

Siegal, B. S. and A. F. H. Goetz 1977. Effects of vegetation on rock and soil type discrimination. *Photogramm. Engng* **43**, 191–6.

Silva, L. F. 1978. Radiation and instrumentation, an overview. In *Remote sensing: the quantitative approach*, P. H. Swain and S. M. Davis (eds), 21–135. New York: McGraw-Hill.

Skaley, J. E., J. R. Fisher and E. E. Hardy 1977. A colour prediction model for imagery analysis. *Photogramm. Engng* **43**, 43–52.

Sklansky, F. 1973. *Pattern recognition: introduction and foundations.* Stroudsburg, Pa: Dowden, Hutchinson & Ross.

Slater, P. N. 1975. Photographic systems for remote sensing. In *Manual of remote sensing*, R. G. Reeves (ed.), 235–324. Virginia: American Society of Photogrammetry.

Smedes, H. W. 1971. Automatic computer mapping of terrain. *Proc. Int. Workshop on Earth Resources Survey Systems, Washington, DC*, **2**, 345–406.

Smedes, H. W., M. M. Spencer and F. J. Thomson 1971a. Processing of multispectral data and simulation of ERTS data channels to make computer terrain maps of Yellowstone National Park test site. *Proc. 7th Int. Symp. on Remote Sensing of Environment, Ann Arbor, Michigan*, 2073–94.

Smedes, H. W., H. J. Linnerud, L. B. Woodlaver and S. J. Hawks 1971b. Digital computer mapping of terrain by clustering techniques using colour film as a three-band sensor. *Proc. 7th Int. Symp. on Remote Sensing of Environment, Ann Arbor, Michigan*, 2057–71.

Smedes, H. W., H. L. Linnerud, L. B. Laver, M. Y. Su and R. R. Jayroe 1972. Mapping terrain by computer clustering techniques using multispectral scanner data and using colour aerial film. *4th NASA Earth Resources Program Rev.* **3**, 61-1 to 61-30.

Sokal, R. R. and P. H. A. Sneath 1963. *Principles of numerical taxonomy*. San Francisco: Freeman.

Stanley, D. J. 1977. Texture analysis in real life – an examination of the efficacy of selected digital methods. In *Proc. Br. Pattern Recognition Assn and Remote Sensing Soc. meeting on Texture Analysis, Oxford*, J. O. Thomas and P. G. Davey (eds), 87–108.

Stanley, D. J. 1978. Digital analysis of radar imagery for vegetation detection in Nigeria. In *Remote sensing in developing countries*, W. G. Collins and J. L. van Genderen (eds), 63–74. Remote Sensing Society.

Steiner, D. 1970. Automation in photo-interpretation. *Geoforum* **2**, 75–88.

Steiner, D., K. Braumberger and H. Maurer 1969. Computer processing and classification of multivariate information from remote sensing imagery. *Proc. 6th Int. Symp. on Remote Sensing of Environment, Ann Arbor, Michigan*, 895–907.

Steiner, D. and H. Maurer 1968. Development of a quantitative semi-automatic system for the photo-identification of terrain cover-types. *Seminar on Air Photo-interpretation in the Development of Canada, Ottawa*, 115–21. Ottawa: Queens Printer.

Steiner, D. and A. E. Salerno 1975. Remote sensor data systems, processing and management. In *Manual of remote sensing*, R. G. Reeves (ed.), 611–803. Virginia: American Society of Photogrammetry.

Strahler, A. H. 1980. The use of prior probabilities in maximum likelihood classification of remotely sensed data. *Remote Sensing of Environment* **9**, (in press).

Strahler, A. H., T. L. Logan and N. A. Bryant 1978. Improving forest cover classification accuracy from Landsat by incorporating topographic information. *Proc. 12th Int. Symp. on Remote Sensing of Environment, Ann Arbor, Michigan*, 927–42.

Su, M. Y. and R. E. Cummings 1972. An unsupervised classification technique for multispectral remote sensing data. *Proc. 8th Int. Symp. on Remote Sensing of Environment, Ann Arbor, Michigan*, 861–79.

Swain, P. H. 1978. Fundamentals of pattern recognition in remote sensing. In *Remote sensing: the quantitative approach*, P. H. Swain and S. M. Davis (eds), 136–87. New York: McGraw-Hill.

Swain, P. H. and S. M. Davis 1978. *Remote sensing: the quantitative approach*, New York: McGraw-Hill.

Swain, P. H. and H. Hausk 1977. The decision tree classifier: design and potential. *IEEE Trans. on Geoscience Electronics* **GE-15**, 142–7.

Swain, P. H., H. Siegel and B. W. Smith 1979. A method for classifying multispectral remote sensing data using context. *Proc. 1979 Machine Processing of Remotely Sensed Data Symp., Purdue University, Indiana*, 343–52.

Tapper, G. O. and R. W. Pease 1973. ERTS-1 image enhancement by optically combining density slices. *Symp. on Significant Results obtained from the Earth Resources Technology Satellite-1*, **1B**, NASA, SP-327, Goddard Space Flight Center, 1179–85.

Tator, B. A. 1960. Photo-interpretation in geology. In *Manual of photo-interpretation*, 169–342. Washington, DC: American Society of Photogrammetry.

Taylor, M. M. 1974. Principal components colour display of ERTS imagery. *3rd Earth Resources Technology Satellite-1 Symp.* Vol. 1: *Technical Presentations Section B*, 1877–97.

Thie, J., C. Tarnocai, G. E. Mills and S. J. Kristof 1974. A rapid resource inventory for

Canada's north by means of satellite and airborne remote sensing. *Proc. 2nd Canad. Symp.*
on Remote Sensing, Guelph, Canada, 200–15.

Thomas, J. O. 1977. Texture analysis in imagery processing. In *Proc. Brit. Pattern Recognition Assn and Remote Sensing Soc. meeting on Texture Analysis, Oxford*, J. O. Thomas and P. G. Davey (eds), 1–25.

Tobler, W. O. 1969. An analysis of a digitized surface. In *A study of land type*, C. M. Davis (ed.), 59–76. University of Michigan, ORA Proj. 08055.

Tou, J. T. and R. C. Gonzalez 1974. *Pattern recognition principles*. Reading, Mass.: Addison-Wesley.

Townshend, J. R. G. 1977. A framework for examining the role of remote sensing in monitoring the Earth's environment. In *Monitoring environmental change by remote sensing*, J. L. van Genderen and W. G. Collins (eds), 1–5. Remote Sensing Society.

Townshend, J. R. G. and C. O. Justice 1980. Unsupervised classification of MSS Landsat data for mapping vegetation in an area of complex terrain: principles and problems. *Int. J. Remote Sensing* **1**, 105–20.

Tunstall, K. W. 1975. Recognizing patterns: are there processes that precede feature analysis? *Pattern Recognition* **7**, 95–106.

Ulaby, F. T. and J. McNaughton 1975. Classification of physiography from ERTS imagery. *Photogramm. Engng* **4**, 1019–27.

VanderBrug, G. J. 1976. Line detection in satellite imagery. *IEEE Trans on Geosci. Electron.* **GE-14**, 37–44.

van Genderen, J. L. 1974. An evaluation of stereoscopic viewing of ERTS and Skylab images. *Eur. Earth Resources Satellite Experiments, Proc. Symp., Frascati, Italy*, 47–55.

Verstappen, H. Th. 1974. On quantitative image analysis and the study of terrain. *Proc. Symp. on Remote Sensing and Photo-interpretation* **II**, 647–65. Banff, Alberta, Canada: International Society for Photogrammetry, Commission VII.

Verstappen, H. Th. 1977. *Remote sensing in geomorphology*. Amsterdam: Elsevier.

Viktorov, S. V., E. A. Vostokova and A. G. Chikishev 1973. Investigations of landscape indicators. In *Landscape indicators*, A. G. Chikishev (ed.), 1–6. New York: Consultants Bureau.

Vincent, R. K. 1972. An ERTS multispectral scanner experiment for mapping iron compounds. *Proc. 8th Int. Symp. on Remote Sensing of Environment, Ann Arbor, Michigan*, 1239–47.

Vincent, R. K. 1973. Spectral ratio imaging methods for geological remote sensing from aircraft and satellites. *Proc. Symp. on Management and Utilization of Remote Sensing Data*, 377–98. Washington, DC: American Society of Photogrammetry.

Vincent, R. K. 1975. The potential role of thermal infra-red multispectral scanners, in geological remote sensing. *Proc. IEEE* **63**, 137–47.

Vink, A. P. A. 1964. Aerial photographs and the soil sciences. *Proc. Conf. on Aerial Surveys and Integrated Studies, Toulouse*, 81–141. Unesco.

Vink, A., H. Verstappen and P. Boon 1965. *Some methodological problems in interpretation of aerial photographs for natural resources surveys*. ITC Publ. series B, no. 32.

Vostokova, E. A. 1973. Recognition and evaluation of indicators. In *Landscape indicators*, A. G. Chikishev (ed.), 7–18. New York: Consultants Bureau.

Wagner, T. W., R. Dillman and F. Thompson 1973. Remote identification of soil conditions with ratioed multispectral data. In *Remote sensing of Earth resources* **2**, F. Shahrokhi (ed.), 721–38. University of Tennessee, Tullahoma.

Watson, R. D. and L. C. Rowan 1971. Automated geologic mapping using rock reflectance. *Proc. 7th Int. Symp. on Remote Sensing of Environment, Ann Arbor, Michigan*, 2043–53.

Way, D. S. 1973. *Terrain analysis*. Stroudsburg, Pa.: Dowden, Hutchinson & Ross.

Weismiller, R. A., S. J. Kristof, D. K. Scholz, P. E. Anuta and S. A. Momin 1977. Change detection in coastal zone environments. *Photogramm. Engng* **43**, 1533–9.

Weismiller, R. A., I. D. Persinger and O. L. Montgomery 1977. Soil inventory from digital analysis of satellite, scanner and topographic data. *J. Soil Sci. Soc. Am.* **41**, 1160–70.

Welch, J. R. and K. G. Salter 1971. A context algorithm for pattern recognition and image interpretation. *IEEE Trans Syst., Man Cybern.* **SMC-1**, 24–30.

Weszka, J. S., C. R. Dyer and A. Rosenfeld 1976. A comparative study of texture measures for terrain classification. *IEEE Trans. Syst., Man Cybern.* **S MC-6,** 269–85.

Wignall, B. 1975. Operator interaction with the Quantimet 720 Image Analyser. In *Remote sensing data processing*, J. L. van Genderen and W. G. Collins (eds), 129–42. Remote Sensing Society.

Williams, D. G. 1975. A system for scanning data from remote sensing applications when recorded on photographic emulsion. In *Remote sensing data processing*, J. L. van Genderen and W. G. Collins (eds), 97–114. Remote Sensing Society.

Yost, E. 1972. Additive colour techniques for the analysis of ERTS imagery. In *Remote sensing of Earth resources* **1**, F. Shahrokhi (ed.), 438–59. University of Tennessee, Tullahoma.

Zinke, P. J. 1960. Photo interpretation in hydrology and watershed management. In *Manual of photographic interpretation*, 539–60. Washington, DC: American Society of Photogrammetry.

5 Regionalization of terrain and remotely sensed data

John R. G. Townshend

5.1 Introduction

Regionalization of terrain has long formed an important component of terrain analysis and continues to be a powerful tool in several distinctive ways. Such regionalization can use data from topographic and thematic maps and ground survey, but historically has relied heavily on conventional black and white air photographs and nowadays increasingly on the newer forms of remote sensing imagery (Peplies & Keuper 1975).

The holistic view of terrain or land (Ch. 1) forms the most common basis for regionalization, namely using the *physiographic or landscape approach* (Section 5.2.1). Despite its subjectivity, the principles of this approach have been utilized (and often developed independently) by many workers, operating in very different parts of the world for a variety of objectives. Moreover the hierarchical structure of such regions is usually explicitly recognized (Section 5.2.2). Areal subdivision by means of air photographs remains an important approach (Section 5.2.3), but subsequently images from space altitudes have become much more important (Section 5.2.4) as well as other forms of imagery (Section 5.2.5) especially radar.

An alternative to the physiographic approach is to use completely objective criteria, the most readily implemented of which are probably morphometric ones (Section 5.3.1). Alternatively regular grids or basic picture elements (pixels) can be used to define uniformly sized cells as the elemental building blocks of regions (Section 5.3.2). Grouping of these units by various clustering algorithms allows regions to be defined objectively. Computer-assisted methods have lessened the key analytical role of regionalization in recent years, but even automated geographic information systems still contain data planes of regionalized data and are required to provide output in regionalized forms (Section 5.4).

Regionalization of the land surface either by subjective or objective methods can serve several different roles, though frequently these are not properly distinguished. This is unfortunate since the merits of a given set of regions are not necessarily the same for each role. Firstly regionalization can help provide an effective summary of an area's characteristics, which often forms a valuable preliminary stage in the familiarization of workers with an area. These regions can then form a useful framework in the planning of a survey. In particular, the design of ground sampling schemes and thus the efficiency of ground data collection should benefit considerably. Regionalization, especially by skilled interpreters, can produce functional units, in which the various terrain components are genetically linked. On this basis, not only can relationships between the immediately observable properties be firmly established, but also inferences made about terrain properties which are not directly observable. Regionalization has therefore an analytical role in terrain analysis. Inherent within the delimitation of such regions, is that relationships will vary between them. Thus regional boundaries have been used to establish the limits

109

of extrapolation of relationships between ground properties using remotely sensed data and thus have improved the efficiency of land cover data collection (Section 5.5). Finally regionalization has a role in improving the display and communication of land characteristics and/or suitabilities. Even coarse resolution images from space altitudes such as Landsat can produce information in vastly too much detail for many planning purposes. Systematic grouping of areas to produce manageable units is often necessary, whether or not regionalization has been carried out at earlier stages.

Regionalization cannot be carried out with the same degree of ease in different parts of the world. But wherever and however they are defined they represent only a way of modelling the diversity of the Earth's surface (Grigg 1967). The usefulness of regions should be judged only according to their success in fulfilling one of the specific roles that we have described.

5.2 The physiographic approach

5.2.1 *Principles*

Physiography is the comprehensive study of surface form, geology, climate, soils, water and vegetation and their inter-relationships.

The physiographic approach to regionalization is thus the subdivision and characterization of the Earth's surface in terms of a broad set of properties. Also at times called the 'landscape approach' its development is founded on the regional concepts of the early 20th century (Mitchell 1973) such as the world subdivision of Herbertson (1905), the physiographic subdivisions of the United States by Fenneman (1916) and Joerg (1914) and the 'topographic types' of Bowman (1914). Bourne (1931), involved in forestry stock-taking procedures, recognized that larger regions could usefully be subdivided into sites which were uniform in terms of land attributes. In Michigan, Veatch (1933) produced a land classification based on soil, vegetation, drainage, and topography, the resultant 'natural land types' being similar to the land system described in the next section. Apparently somewhat independently, Milne (1935) developed his concept of the catena in East Africa, where soils were mapped in groupings related to topographic position.

Post-war development of the physiographic approach has been especially strong in the reconnaissance of several areas of Australia by the CSIRO (Commonwealth, Scientific and Industrial Research Organization) (e.g. Christian & Stewart 1953). Similarities of this work with that of workers in South Africa and the United Kingdom, led to a joint meeting and agreement on a common system of nomenclature (Brink *et al.* 1966). Other notable examples of similar work include the Canadian biophysical land classification (Canada Land Inventory 1970; Gimbarzevsky 1978), and the schemes of several Russian workers (e.g. Vinogradov *et al.* 1962, Chikishev 1973).

The relative importance of the physiographic properties chosen will vary between surveys, depending on the type of area under investigation, and the ultimate use of the results. The former is self-evident if one considers the relative importance of characteristics in making a physiographic description of a semi-arid plain compared with a humid tropical mountain region. Choice of variables is also dependent on assumed relationships between land characteristics (Section 5.2.3).

In many operational surveys using the physiographic approach, a bias towards geomorphological terms is readily discerned. An extreme view is that of Solntsev (1962) who writes that a

> . . . geologic–geomorphic foundation is always the principal factor in segregating landscapes, and a genetic system of land classification must therefore be based on it.

A counterbalance to this bias is provided by ideas of landscape ecology and biocenology (Moss 1969, Troll 1971), whose origins lie in the biological sciences. For example Tansley (1935) writes that

> The ecosystem consists of both organic and inorganic components which may be conveniently grouped under the heads of climate, physiography and soil, animals and plants,

a definition clearly not very different from that of land given earlier, though Tansley uses 'physiography' in a restricted geomorphological sense.

In practice, biotic factors are of relatively much greater importance in an ecological approach than the 'physiographic' one we have outlined. Moss (1969) has objected specifically to the use of physiographic 'land systems' methodology for the appraisal of agricultural development in the humid Tropics arguing that a biocenological one is more relevant, in which present-day land use is taken fully into account. Stress should be laid

> on those properties which are reciprocal and interacting, and those which are essentially environmental and unidirectional. (Moss 1978, p. 23.)

Clearly this approach is potentially much more dynamic in emphasizing flows of energy and nutrients, but in practice the actual mapped units will often be similar to those obtained utilizing the land systems approach (Cooke & Doornkamp 1974, p. 345) though there will be differences in emphasis on the type of data which are collected.

We would not invariably support or oppose the views of either Solntsev or Moss or other workers arguing for the pre-eminence of particular sets of data. The specific information requirements of particular surveys will strongly influence the choice of land characteristics which need to be described. We agree with Vink (1975) that 'artifactial' or man-made features in the landscape should also be incorporated in the description of terrain type. Many landscapes exist where the most dominant topographic features are man-made, as in the polders of Holland or fenlands of eastern England or in heavily terraced areas. Similarly it makes little sense to describe the hydrological properties of heavy clay lands without reference to existing under-draining.

5.2.2 *The hierarchical structure of terrain*

It is usually assumed that terrain is composed of units,[1] variously called elements or sites, which are considered internally homogeneous. Although this can never be

---

[1] We follow Mitchell (1973) in using the term 'land unit' or 'terrain unit' to refer to an areal subdivision of terrain of any size or order (cf. Brink *et al.* 1966, Christian & Stewart 1968, and Chapter 7 of this book by D. F. Williams).

completely true, it often forms a useful approximation and conceptualization of reality. In practice such 'atomic' components of the landscape are usually much too small to be mapped separately for any but the most detailed surveys, and the smallest areal unit ever mapped is usually the *facet*, which consists of one or more elements, but which nevertheless is still reasonably homogeneous (Mitchell 1973). In most physiographic surveys, even facets are not separately mapped but instead the smallest mapping unit is formed of distinctive assemblages of facets. These are most commonly called 'land systems' using the terminology of Australian CSIRO Division of Land Research and Regional Survey but have been called 'natural regions' or 'terrain systems'. More formally a land system is an area with 'a recurring pattern of topography, soils and vegetation' (Christian & Stewart 1968, p. 242), with a corresponding pattern of hydrology and geology and is thus closely related to the concept of land described in Section 1.2. Land systems are thus not necessarily internally homogeneous but are areas with a characteristic recurrent heterogeneity.

The ecoclass system of the US Department of Agriculture Forest Service links terrestrial and aquatic ecosystems (Buttery 1978). *Ecological land units* (ELUs) are defined by distinctive combinations of vegetation 'systems' and 'land systems' based on soils, geology, land shape and climate and *ecological water units* (EWUs) by distinctive combinations of 'land systems' and 'aquatic systems'. This has subsequently been developed to include a soil system as well, to define modified ELUs and EWUs (Fig. 5.1), the former corresponding closely to the land systems of the CSIRO. For the sake of clarity it should be pointed out that in the ecoclass terminology the word 'system' refers to the whole hierarchy of areal units whereas in the CSIRO system the term refers only to one level of the hierarchy.

Stemming from the hierarchical regional systems of geographers such as Fenneman (1916), Wooldridge (1932), Unstead (1933) and Linton (1951), physiographic schemes often have additional levels of areal aggregation, and although it is impossible to be sure about precise correspondence of different terms, Table 5.1 shows the believed equivalence of several schemes.

If assemblages of units at one level in the hierarchy are mapped rather than individual lower order units themselves, this can undoubtedly reduce time and costs substantially, and if working at a reconnaissance scale it can produce satisfactory results. But the very great differences between facets within land systems can restrict

Figure 5.1 The ecoclass system (from Buttery 1978).

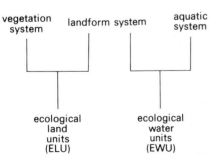

Initial ecoclass system

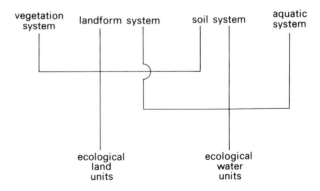

Modified ecoclass system

Table 5.1 Approximate equivalence of areal subdivisions of terrain.

Bourne (1931)	Unstead (1933)	Isacenko (1965)	MEXE (1965)	Grant & Aitchinson (1965)	MEXE/ CSIRO Brink et al. (1966)	CSIRO Christian & Stewart (1968)
site	feature	facies	land element	land component	land element	land element
region	stow	urocise	land facet	terrain unit	land facet	land unit
	tract	mestnost	recurrent land pattern	terrain pattern	land system	land system
	tract group	okrug unterprovinz	land region / land province	province	land region / land province	
					land division / land zone	

the usefulness of maps which fail to depict them individually. Vakilian and Mahler (1974) for example whilst recognizing the need for a multipurpose land evaluation survey in Iran rejected the use of land systems as a mapping unit, since this would have led to the grouping of lands of quite different potentialities, in part as a result of the mountainous terrain of much of Iran which creates an intricate pattern of land facets. Nevertheless where neighbouring facets did have similar characteristics, they were grouped and mapped together.

If only land systems (or more complex areal units) are mapped, the procedure for using the results becomes two-stage (Beckett 1968) requiring the user himself to locate the homogeneous areas within the land systems before the results can be used. This sets a limit to the usefulness of the results since as Beckett (1968, p.54) points out in a slightly different context,

> It must not be forgotten that the presumed user of the maps turns to it for information in the expectation that he can thereby obtain the information with less effort than if he went out into the field to collect it himself.

The description of physiographic regions is enhanced by means of block diagrams, in which the relative spatial and vertical locations of their components are depicted. These are especially useful for indicating the position of facets within land systems (Fig. 5.2), and thus aid the location of unmapped units.

Such block diagrams do not necessarily correspond to any one part of an area, but model the relationships, both geographic and genetic, between the units. When produced before field survey they represent a series of hypothetical statements about the terrain which need to be verified. Their principal difficulty of application is in the representation of soils, since the vertical variation of pedological properties are usually small relative to the overall relief variation and thus cannot readily be

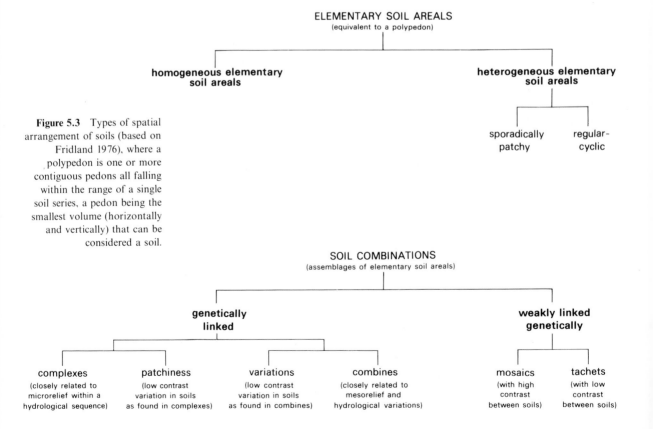

Figure 5.2 Block diagram showing sub-divisions within a land system (from Perrin & Mitchell 1969). UK Crown Copyright: Ministry of Defence.

represented. Principles of the construction of block diagrams in general, may be found in Monkhouse and Wilkinson (1963).

Even with a map of land systems, block diagrams and tables, locating a facet in the field may prove difficult. For example, where the units are described as 'dissected alluvial fans' or 'colluvial foot slopes' only those with some geomorphological training will be able actually to locate them precisely on the ground. Precise

Figure 5.3 Types of spatial arrangement of soils (based on Fridland 1976), where a polypedon is one or more contiguous pedons all falling within the range of a single soil series, a pedon being the smallest volume (horizontally and vertically) that can be considered a soil.

geographical location must be a *sine qua non* for useful planning units of even a reconnaissance type and the fact that this is often not provided by land systems surveys undoubtedly has led to them being under-utilized.

Although there is a consensus of opinion about the hierarchical structure of land, there are sharp disagreements about the level in the hierarchy at which investigation should start. Many workers believe that the land system or even the land province forms an appropriate starting level (e.g. Christian & Stewart 1968, Mitchell 1973). The contrary view that the simplest elements should be mapped first has been espoused especially by Wright (1972), and similar views are expressed by Viktorov (1973).

Surprisingly, relatively little has been done on the various types of spatial arrangement at various levels of the hierarchy. A notable example in the context of soils is that of Fridland (1976) described in outline form in Figure 5.3.

5.2.3 *Visual regionalization using air photographs*

An essential precursor for the regionalization of areas represented on air photographs is experience in their interpretation. Principles of interpretation are found in several standard texts (e.g. American Society of Photogrammetry 1960, Leuder 1959, Lo 1976), and a hierarchical scheme of interpretation was suggested in Section 4.7.

Distinctive assemblages of the levels of observation and interpretation, described in Section 4.7 and Figures 4.18 and 4.19, can be used to create regions of different kinds (Fig. 5.4). Hence if we use only the directly observable properties of the image without attempting any interpretation, *photomorphic regions* are produced. Where an image is found difficult to interpret, such a regionalization can be very helpful in simplifying the image and thus enables the primary interpreted properties to be grouped in more manageable assemblages and to provide a rational basis for ground sampling to discover what they represent (MacPhail 1971). Difficulties in interpretation may arise due to lack of overall experience of the interpreter or unfamiliarity with the kind of terrain or type of image. Experienced interpreters working in familiar terrain will have great difficulty in carrying out a purely photomorphic regionalization since recognition of land characteristics is almost instinctive, but occasional areas which cannot be interpreted with confidence may be found and should be separately delimited so they receive more intensive ground work.

Using either primary, secondary or tertiary interpreted properties, it is possible to derive either single or multiple property regions (Fig. 5.4). Their recognition in turn may help in the identification of other primary, secondary or tertiary properties. For example, a map of landforms can help in the location of both surface and subsurface hydrological characteristics. It is also possible directly to combine single property regions to produce multiple property regions and consideration of the interactions between primary, secondary and tertiary properties can be used to produce comprehensive physiographic regions. This stage need not necessarily have been preceded by other regionalizations. The traditional land systems method then uses the physiographic regions along with ancillary data, including block diagrams, to make inferences about land suitabilities (quaternary interpreted properties) and hence to produce land suitability regions. Not surprisingly, adopting this procedure, the boundaries of the latter will be similar to those of the physiographic regions.

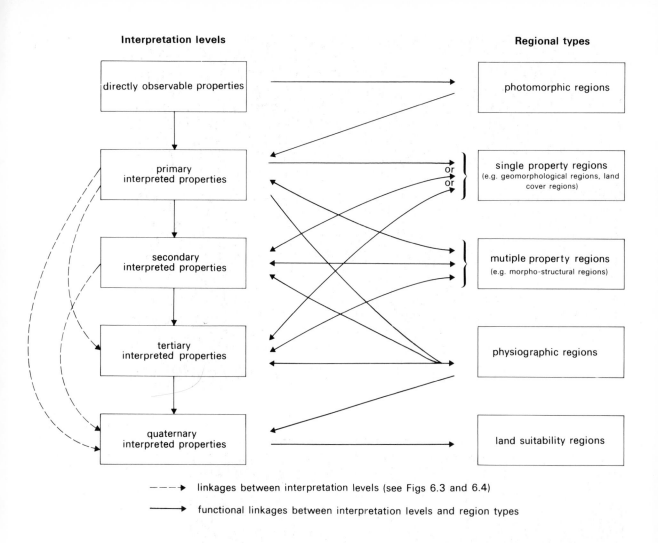

Interpretation levels

Regional types

directly observable properties → photomorphic regions

primary interpreted properties

single property regions (e.g. geomorphological regions, land cover regions)

secondary interpreted properties

mutiple property regions (e.g. morpho-structural regions)

tertiary interpreted properties

physiographic regions

quaternary interpreted properties → land suitability regions

- - - → linkages between interpretation levels (see Figs 6.3 and 6.4)

⟶ functional linkages between interpretation levels and region types

Figure 5.4 Relationship between interpretation levels and regionalization.

From Figure 5.4 it can be seen that land suitability regions might alternatively be derived only as a final stage following the various states of interpretation on the left, without regionalization at any previous stage. Clearly many different route-ways can be followed through this scheme with several different end-points.

The ease with which regionalization can be executed depends on the type of terrain being studied. Figure 5.5 contains regions which probably nine out of ten interpreters would recognize. In Figure 5.6, although clear variability is apparent there is no clear boundary which can be located between the regions. If a map of regions is ultimately required, the broad zones of change should be subject to more intensive ground data collection. If no sensible regions can be perceived at all, then there may be no option but to lay out a series of ground sample points, probably in a regular grid arrangement. Figures 9.3 to 9.5 provide examples of the visual recognition of air photographs.

The above discussion is made with particular reference to conventional black and white aerial photographs, but is easily extended to true colour and false colour aerial photographs, by taking into account the additional directly observable properties.

116

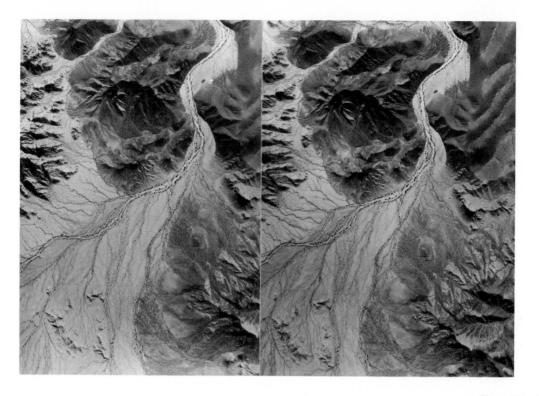

Figure 5.5 Air photographs easily divisible into regions especially when viewed with a pocket stereoscope (location: south Arizona). Area depicted approximately 2 × 3 km. US Geological Survey photographs.

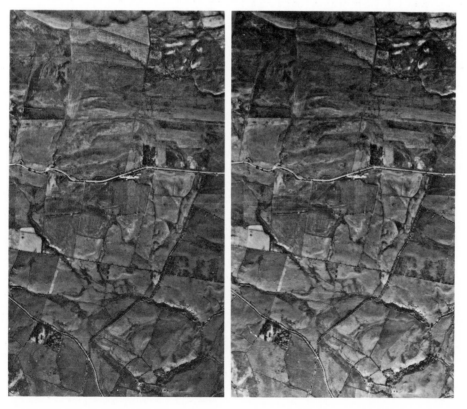

Figure 5.6 Air photographs of an area with different terrain types, especially noticeable when viewed with a pocket stereoscope, but which are difficult to separate with boundaries (location: near Gell, Clywd, North Wales). Area depicted: approximately 2·5 × 3·5 km. UK Crown Copyright, Department of Environment.

5.2.4 *Visual interpretation of images obtained from space altitudes*

The unfamiliarity in appearance of many areas from space altitudes means that recognition of image pattern regions akin to photomorphic regions (Fig. 5.5) will be necessary more frequently than is the case with conventional air photographs. Usually the stereoscopic model will be unavailable or at best present in a much weaker form than that obtainable from conventional air photographs and thus reliance will be placed more strongly on grey tone or colour variation. If images with relatively low Sun angle are used, the amount of morphological information will be much higher, but conversely the consistency with which land cover information can be recognized will be lower. Consequently, the availability of images from more than one season can be of great value in recognizing physiographic regions. The principal limitation of such images is that they contain fewer data than images obtained from lower levels, but paradoxically this results in one of their principal virtues as well: the overview and greater degree of generalization of such images, means that the overall subdivisions are often easier to recognize.

The growing familiarity of interpreters with such images means that reliable identification of terrain features is increasingly possible. Nevertheless consultation

Figure 5.7 Multi-level sampling based on space images.

Figure 5.8 (*opposite*) Image from a space platform with easily recognised regions. It depicts an area around Ksar es Souk, Morocco, with the Atlas fold mountains in the north, extensive plains of alluvial fans and stone pavements immediately to the south and finally highly eroded older fold mountains in the lower part of the image. These can in turn be readily sub-divided into smaller units some of which may be identifiable at least provisionally. Area depicted: 185 × 185 km. NASA image E-2402-10082-7.

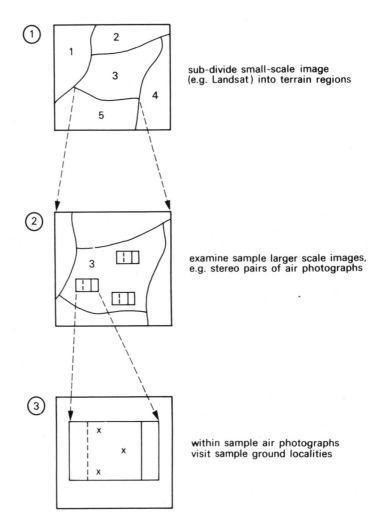

sub-divide small-scale image (e.g. Landsat) into terrain regions

examine sample larger scale images, e.g. stereo pairs of air photographs

within sample air photographs visit sample ground localities

118

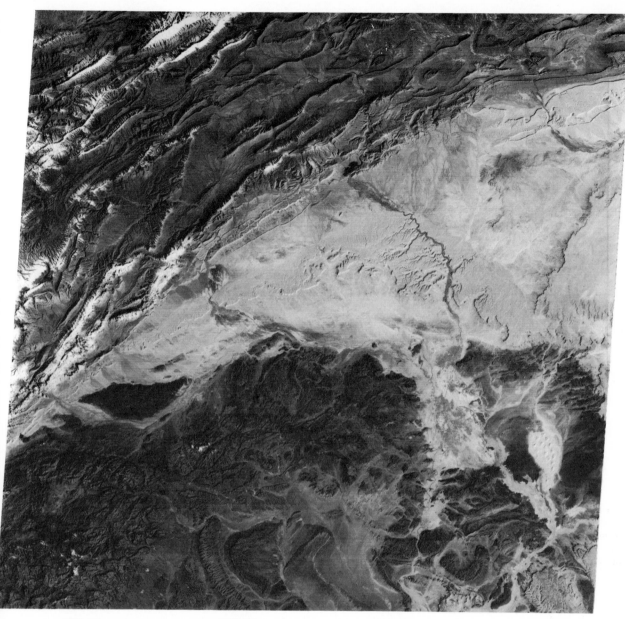

W005-301   W005-001   W004-301   W004-001
28FEB76 C N31-39/W004-38 N N31-38/W004-34 MSS   7 R SUN EL37 AZ133 190-5601-G-1-N-D-1L NASA ERTS E-2402-10082-7 01

of sample stereo pairs of air photographs, even from a much earlier date, will facilitate the task of interpretation. Thus a multilevel sampling scheme can be derived such as that shown in Figure 5.7, and the time taken in field observations should be substantially reduced and/or made much more effective. Various examples of this approach have been reported. Peplies and Keuper (1975) describe the definition of photomorphic units in the San Joaquin Valley using Landsat images and Brueck (1976) defines geobotanical units using similar methods in a study of forage production in Iran. Other applications of Landsat images include that reported by Gimbarzevsky (1974) in the recognition of 'biophysical' regions in two parts of Canada and in Chapter 8 of this book the role of physiographic regions in various arid and semi-arid areas is outlined. Story *et al.* (1976) report that only 119

Figure 5.9 Image from a space platform with less easily recognizable regions than in Figure 5.8. Black and white rendition of a false colour infrared photograph taken from a high altitude Skylark earth resources rocket of the pampas of Argentina near Rio Cuarto. See text for description. The first stage of the rocket is visible, in this high angle oblique, which depicts an area approximately 60 km across at the centre of the image.

50% of the detail of an original land system subdivision of part of Australia using air photographs could be obtained from Landsat images but the methods of image processing and analysis were relatively simple involving the use only of black and white prints and transparencies viewed with a $10 \times$ magnifier.

The ease with which regions can be recognized varies enormously between different terrain types, as was shown to be the case with conventional air photographs (Figs 5.5, 5.6). Figure 5.8 illustrates an area of Morocco near Ksar es Souk. Many of the boundaries recognizable are self-evidently ones which virtually any interpreter would choose. In general, areas where vegetation cover is low or natural and semi-natural vegetation is still present will be regionalized relatively easily. An exception to this is in tropical rain forests observed using Landsat images, which without enhancement usually appear almost uniform in tone except for open water bodies. Examples of the regionalization of Landsat images can be found in Figures 8.9, 8.10 and 9.6.

Where the land has been cultivated, the resultant field pattern tends to dominate the appearance of the area and recognition of terrain regions becomes much less clear, particularly if the field size is large compared with the resolution of the images. When conventional air photographs are used the latter can hinder interpretation, but this disadvantage is counterbalanced by the availability of the stereoscopic mode. With small-scale space-altitude images, subdivision of agricultural areas can be much more difficult. Figure 5.9 shows a part of the humid pampas of Argentina, taken with a camera in a Skylark rocket. Image-pattern regions are perceptible utilizing field size and shape patterns and tone patterns. The former must be treated carefully because field size and shape undoubtedly are a result of cultural factors as well, though the influence of the latter is unequivocally displayed in very sharp boundaries between some of the patterns. The most obvious image regions are a series of broad linear features. These have been found to represent current flood plains as well as palaeo-flood plains unoccupied by major contemporary rivers. They are identifiable partially by their distinctive field patterns and partially by the higher amount of near-infrared radiation reflected from the crops growing in the flood plains, as a result of higher amounts of available moisture. In terms of land suitability for crops these physiographic units are quite distinctive (Savigear *et al.* 1975) and by analogy with nearby areas probably have considerable significance as fresh water aquifers (Kruck & Kantor 1975).

5.25 *Visual interpretation of other forms of remotely sensed images*

The principles outlined in the previous two subsections can be applied to any type of remotely sensed images. Those described for conventional air photographs are readily extendible to high-altitude aircraft photographs (Peplies & Keuper 1975), such as the false colour infrared photographs now available for much of the U S and also for some other countries such as Guatemala and Sierra Leone.

Regionalization procedures have also been found fruitful for the analysis of radar images, where the interactions between terrain properties and radiation are much less well understood than in the visible and near-infrared parts of the electromagnetic spectrum. Nunnally (1969), using K-band images (H H and H V polarizations) of the Asheville Basin of North Carolina, found that by outlining 'image variation in tone, texture, pattern and shape' varying associations of physical and cultural phenomena could be delimited, which he called 'integrated landscape units'. Lewis (1971) also using K-band imagery in a geomorphic evaluation of radar imagery of south-eastern Panama and north-western Colombia defined geomorphic regions using similar criteria to Nunnally.

5.3 Objective definition of regions

Regionalization of images by visual interpretation, which was described in the previous section, inevitably involves subjective decisions, and there is a natural tendency therefore to search for more repeatable, objective procedures. A number of alternatives are available, the first group of which involves the use of morphometric units (Section 5.3.1) defined either from remotely sensed data or more frequently from topographic maps. Alternatively, terrain properties can be derived for regularly shaped areas usually grid squares which can then be grouped by various algorithms to create regions (Section 5.3.2).

5.3.1 *Morphometric units*

The most frequently used type of morphometric subdivision of terrain is the drainage basin or catchment. In order to compare like with like, various methods of stream and hence drainage basin labelling have been derived, usually by means of their topological properties. The most commonly applied is the Strahler (1964) ordering system (Fig. 5.10), though several alternatives exist (Haggett & Chorley 1969). The principal advantage of drainage basins as terrain regions is that they are functional units within which many land properties and processes have distinctive spatial arrangements (Gregory & Walling 1973). The resultant regions are thus comprised of the drainage basins with the corresponding orders of the streams to which they contribute.

Although apparently an attractive system, the drainage basin has many limitations as a practical spatially comprehensive unit for storage of land information. Because of the hierarchical structure of fluvial networks, catchments of a particular order cannot completely occupy an area (Fig. 5.10) and any given locality may be within basins of more than one order. Furthermore the scale and resolution of the data source has a strong effect on the number of streams detected and hence on the order of the drainage basins: thus the fluvial network can be defined in several different ways (Gregory & Walling 1973). For many purposes, the boundaries of the units are at inappropriate localities, since the land on either side of the watershed is likely to have similar suitabilities for many uses.

Comparable sorts of schemes can be derived for other types of terrain, notably karst scenery developed on limestone characterized by sink holes (Williams 1972).

Figure 5.10 Ordering of river catchments using the Strahler (1964) system.

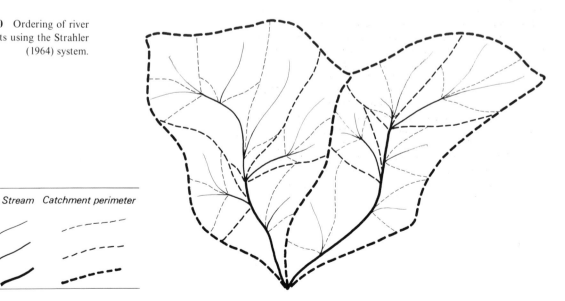

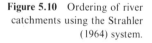

Order	Stream	Catchment perimeter
1		
2		
3		

Plate 1 A false colour infrared (FCIR) photograph taken from an aircraft in September 1977. The original was a 23×23 cm transparency on Kodak 8443 film, and the ground area depicted is approximately $1 \cdot 6 \times 1 \cdot 6$ km, near Reading, England. Some prominent land cover types are indicated by letters as follows: A, grass (natural pasture); B, maize (corn); C, wheat, ready for harvest; D, fields from which grain crops have recently been harvested (stubble); E, kale; F, broadleaf woodland. Simultaneous thermal imagery of the same area is shown in Figure 2.3, and multispectral imagery in Figure 2.4.

Plate 2 (a) Density slice of side-looking airborne radar image (Fig. 2.7) in which the whole range of tones has been equally sub-divided. (b) Density slice of (a) in which the lightest tones only have been sliced.

Plate 3 A Landsat colour composite of the Bay area, California, produced from channels 4, 5, and 7 depicted in Figure 2.5. Channel 7 is represented by red, channel 5 by green and channel 4 by blue colours. The relative importance of the channels can be inferred from the colours in the plate. For example, the bright reds of the redwood forests near the coast, indicate considerable absorption in channels 4 and 5 compared with channel 7. Yellow colours are caused by relatively high reflectance in channels 4 and 5 and relatively low reflectance in channel 7. A comparison of the separate bands depicted in Figure 2.5 with the colour composite demonstrates the value of the latter in providing an overview for visual interpretation. We can at once recognise the main terrain regions. The folded coast mountains of Mesozoic and Tertiary rocks can be readily distinguished from the Sacramento Valley to the north east infilled with Quaternary sediments. Small areas of the latter can also be seen in valleys within the mountains. Several active faults cross the area, in particular the San Andreas Fault bounding the eastern flanks of the coastal mountains, with their redwood forests. These contrast with the browns and oranges of Californian oaks and evergreens with chapparal on the mountain range to the immediate north east, which in turn is readily distinguishable from the checkerboard field pattern of the irrigated Sacramento Valley. Central urban areas can be recognised by their blue colour and fine texture. Suburban areas with pinker hues, because of the higher proportions of vegetative cover are more difficult accurately to delineate (NASA image E-1075-18173). Area depicted: 185×185 km.

Plate 1

(a)

Plate 2

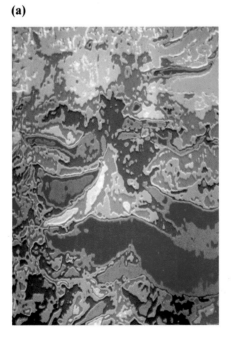

(b)

Plate 3

Plate 4

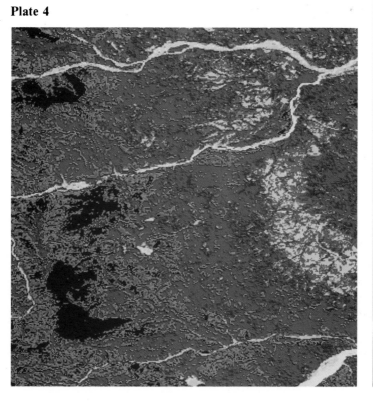

Plate 5

Plate 4 A classification of the land cover types in the central Agri and Sinni catchments in Basilicata Province, southern Italy. The classes were derived using an unsupervised classification of ratioed Landsat data (see Sections 4.10 and 6.3.6 for explanation of methods). The classes are as follows:

(1) White	Bare river gravels.
(2) Sand	Bare surfaces with small proportions of vegetation (less than 10%) including both eroded clays and river alluvium.
(3) Aqua	Bare surfaces with shrubs and small amounts of other vegetation. Percentage vegetation cover is more than 10% and less than 35%.
(4) Tan	Herbaceous vegetation: predominantly agricultural land with large proportions of bare ground (up to 65%). Less than 5% of the cover consists of trees and shrubs.
(5) Dark grey	Herbaceous vegetation, predominantly various grades of pasture with small amounts of bare ground. Less than 5% cover of trees and shrubs.
(6) Red	Herbaceous vegetation, predominantly pasture, with shrubs and trees covering more than 5% but less than 15% of the ground. Bare ground is less than 5%.
(7) Light blue	Herbaceous vegetation with shrubs and trees covering up to 45% of the ground.
(8) Light green	Open woodland, with up to 80% cover of trees and shrubs and up to 50% herbaceous cover.
(9) Dark green	Closed woodland with over 80% tree cover.

Area depicted: 24 × 24 km.

Plate 5 Colour composite produced from the first three contrast-stretched principle component images derived from a six band multispectral scanner image of part of the Blackwater wildlife refuge, Maryland, USA, obtained from an aircraft. Colours green, red and blue have been assigned to the first three components respectively. The dark blue colours represent areas of open water, the reds salt marsh, and the various shades of green are mixed woodland. The area shown is approximately 1·3 km long.

Instead of using physiographic schemes for collecting and storing data, an alternative is to subdivide an area by a regular grid and thus create an array of regular areal units for these purposes. Usually the regular grid will define square areal units (e.g. Miller & Pearson 1971) though triangles have been used (Hobson 1967 and 1972, Gold 1978). The use of regular units defined by regular grids for the collection and storage of land data has the advantage over physiographic and geomorphological regionalizations of subdividing terrain into units with exactly the same size (Mather 1972) which aids comparison between different areas. On the other hand varying land complexity may demand variable grid sizes. This problem is not solved simply by choosing the smallest required grid size. For example, a fine grid placed over fluvially dissected terrain might yield many cells with zero drainage density simply because grids fall on valley side slopes between actual channels. Equally, choosing larger cells may lead to different types of terrain being circumscribed within the same cell. Choice of the most appropriate cell size remains a difficult and at times intractable problem.

Even with relief data alone, the possible ways of describing an area of land become very large indeed, and the choice of properties must be based on their likely usefulness in the final terrain descriptions and the ease with which the measurements can be made. For example, the apparent attractiveness of land descriptions based on harmonic and spectral analysis (e.g. Stone & Dugundji 1965, Preston 1966, Rayner 1972) are undoubtedly hindered for many practical applications by high computational demands.

The use of grid squares for terrain descriptions received its greatest impetus from the parametric approach for military applications especially from the US Army Engineer Waterways Experiment Station at Vicksberg (USAEWES 1962, 1965, 1968). In the original formulation of this approach (Fig. 5.11) several different characteristics are recorded for each cell. Class boundaries were then set within each data plane and the resultant 'factor maps' overlaid to produce a 'factor complex' map, each resultant region typified by values of each factor: these are mathematically manipulated to produce an estimate of terrain performance. According to Benn and Grabau (1968), the characteristics chosen and the class limits set should be totally dependent on a previously derived mathematical model of the relationships between terrain performance and terrain characteristics.

Schemes have been devised for estimating land capability similar to that outlined in Figure 5.11, but without stage 1 being strictly adhered to, in that readily available sets of land characteristics were used. If successful such an approach is to be preferred since it demands much less special purpose data collection, which can seriously restrict the applicability and usefulness of the parametric method (Aitchison & Brown 1968). Examples are discussed in Section 5.4.

Unsupervised classification of regular arrays of remote sensing pixels and other terrain data stored in cells usually produces regions of quite different kinds from the physiographic ones described in the second section of this chapter, since they are solely dependent on the values of properties at each individual cell rather than on the spatial organization of properties (Haggett 1972) and contiguous assemblages of cells to create large multicelled regions will usually occur much less commonly. This is not necessarily desirable and algorithms such as ECHO (extractions and classification of homogeneous objects) have been derived, which firstly groups adjacent pixels into homogeneous fields and subsequently classifies each field as a unit. Application of this algorithm has been found not only to produce

123

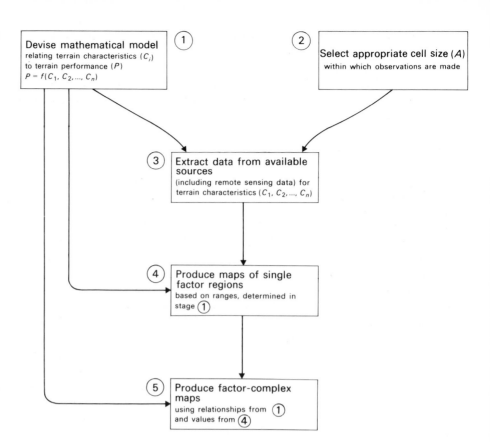

Figure 5.11 Stages in the parametric analysis of terrain (based on Benn & Grabau 1968).

simpler and neater categorizations but has also raised overall successful classification levels (Kettig & Landgrebe 1976). Thus the classification accuracy of a given cell is improved by knowledge of the character of adjacent areas. In other words a type of contextual information is being used (Section 4.13). Use of contextual information is in effect one of the principal ways by which visual physiographic regionalization achieves its success and it is to be expected that improvements in classification by computer-assisted methods will follow the inclusion of more contextual information.

If only a single property is being mapped, such as one particular land characteristic, then creation of satisfactorily comprehensible displays may be produced by choice of appropriate class divisions. This is equivalent to thresholding which is commonly used in digital image processing for image segmentation (Gonzalez & Wintz 1977), and can be extended to several variables using multivariable histograms. Such methods are completely dependent on point properties, but it is also possible to use regional properties. One possible mode of implementation is to use various features, such as point values and/or textural properties, in order to create cluster regions as has commonly been done in geography (Lankford 1969, Mather 1972, Pocock & Wishart 1969, Spence & Taylor 1971). If regions are desired comparable in appearance to those perceived by human interpreters then strong contiguity constraints will be required (Gonzalez & Wintz 1977). The diversity of available techniques and variations in implementation (Tou & Gonzalez 1974) mean that, although each may be internally consistent, an enormous range of regionalizations may be obtained dependent on initial decisions by the human analyst. Thus the results are only objective in a quite restricted sense (Johnson 1968).

Production of physiographic regions themselves by cluster analysis methods is feasible as has been shown by Speight (1968) who, for one study area, produced regions comparable to land systems. On the other hand, land systems containing very dissimilar subunits (for example, low angle pediment slopes with steep mountain slopes above them) are unlikely to be grouped together by cluster analytical methods even if we build in strong contiguity constraints.

An alternative approach is by use of syntactic pattern recognition, which in contrast to the decision-theoretic approach we have previously discussed, has a structure-handling capability, so that the relations between different parts of the image are described (Gonzalez & Thomason 1978). It is assumed that images are composed of primitives; the rules defining the way in which they are assembled are called a 'grammar'. Such methods have been used successfully for example in fingerprint identification. Their application to the study of land, depends on the definition and recognition of primitives within images or digital terrain arrays. Fu (1976), for example, used clusters derived using a Bayes classifier to produce the pattern primitives. Other primitives could include stream segments or even slope segments, though encoding by x, y co-ordinates would be required for successful implementation of the latter (see Section 5.4). Fu (1976) analyzed Landsat imagery of Marion County (Indianapolis), and derived linguistic rules to form cloud-shadow pairs and recognize highways. Results suggested significant improvements using this method over simple point classification. It remains to be seen whether landscapes can be described in syntactic terms with sufficient simplicity to allow their recognition by matching with known landscape types or whether we can build regions by means of similarities in the grammars of their component parts.

5.4 Geographic information systems

The previous section discussed the impact of the arrangement of data in regular arrays of cells (including pixels) on regionalization and the way in which the data can be manipulated in order to provide useful terrain information. In it we mentioned several times the problem of choosing the appropriate cell size. An alternative approach is to encode the data using an x, y co-ordinate system. This is preferable for certain types of data, such as point or line data and for locating boundaries. In particular, if we wish to incorporate information from sources such as geological and soils maps, then polygons defined by sets of x, y co-ordinates will form a more precise representation of the original data.

If we wish to use large volumes of spatial data of several different kinds, it may become desirable to develop an automated geographic information system (GIS). The term 'geographic' in this context simply means that all data are referenced by a set of geographic co-ordinates. The functions of a geographic information system (Knapp & Rider 1979) include the production of the required input data, if unavailable from existing sources. Thus human or machine-assisted classification of remote sensing data may form one of the first stages. Secondly, the data will need to be encoded in a form suitable for handling by a digital computer. Thirdly, a data management subsystem will be required for such procedures as storage, retrieval and updating. Analysis of the data will require an ability to overlay different data planes, change co-ordinate systems, apply statistical algorithms as well as applying algorithms for implementing terrain attribute and resource models to make

125

forecasts of the future distribution of terrain characteristics and to make estimates of land suitabilities. Finally, the results of the analysis will need to be displayed in various forms. Clearly considerable human involvement is necessary but significant steps have been made to automate these procedures.

In a geographic information system we may thus eventually have a large number of data planes, representing several different types of data. These might include remote sensing (possibly from several dates), thematic map data of the area's geology or pedology, topographic data (probably in the form of arrays of digital height data rather than encoded contours), administrative boundaries, drainage basin limits, and so on. The parametric system of land evaluation described in the previous section can be considered as a type of geographic information system. However, the recording of all the data planes in the same format reduces many of the complexities associated with the operation of most geographic information systems. In practice most systems will contain data planes derived from sources with very different formats. It is unfortunately not possible to state unequivocally whether the grid (or cell) data structure is better or worse than the x, y co-ordinate structure. Knapp (1978), in reviewing their relative merits, concludes that the grid system has advantages in terms of ease of data analysis and manipulation in order to carry out such operations as overlaying and hence, in mapping composites, showing the spatial correspondence of different properties and in making estimates of connectivity and the nearness of localities to each other. But it has the disadvantages of tending to aggregate dissimilar characteristics into the same areal unit, reducing the precision of geographic location and creating data which is difficult to store efficiently. The x, y co-ordinate structure has the advantages of allowing high precision of geographic location and permitting high storage efficiency, but it has the major disadvantages of causing the data to be difficult to manipulate and analyze. Thus there is the need to be able to transform the data from one format to another.

In principle, remote sensing data is suitable for incorporation into geographic information systems, but in practice this is achieved only through the solution of considerable practical problems (Knapp & Rider 1979) involved with the remote sensing products themselves (e.g. difficulties of data format registration through geometric correction, variable image quality and at times inadequate resolution) and problems associated with the classes resulting from the analysis of remote sensing data (e.g. inadequate class field data, insufficient classification accuracy, and at times the inappropriateness of the categories of classified data). Nevertheless, many different data planes have been successfully massaged together and have produced useful final results (e.g. Schlesinger *et al.* 1978, Tom & Christenson 1978, Missillati *et al.* 1979). One can either use the remote sensing data to create a classified map, or image, which is incorporated as a plane in the geographic information system, or alternatively the original remote sensing data can be input as multiple planes and advantage taken of ancillary data planes to improve the classification utilizing the remote sensing data. A valuable collection of papers on developments of geographic information systems can be found in the proceedings of a conference held at the Laboratory for Computer Graphics and Spatial Analysis (Harvard University 1979).

Essentially, a geographic information system is a way of efficiently handling unwieldy quantities of geographic data, and the nature of the output is largely a function of the algorithms used in the analysis. Hence we might use such a system to display the distribution of one or more terrain characteristics: for example the distribution of agriculture on soils of different qualities (Cowan *et al.* 1976).

Alternatively, changes in land characteristics from one time period to another could be displayed (e.g. Schlesinger *et al.* 1979). A conceptually more advanced approach is to apply an algorithm using the data available in the geographic information system in order to estimate land suitabilities. An example of one such scheme is that described by Beeman (1978) who uses forest cover, forest stand size, canopy closure, soil productivity, position on slope and proximity to open water to devise a white-tailed deer habitat model. Residential suitability analysis in Minnesota was derived from commuting distances, distances to roads, soil capability to accept septic tanks and the presence or threat of groundwater contamination (Hicks 1977), using information available from the Minnesota Land Information System. Other applications include transmission line siting (Cross 1979), estimating soil erodibility, and the likelihood of lands contributing to water quality degradation (Svatos 1979). For the San Bernadino National Forest, a wildlands recreation study has been carried out (ESRI 1979) in which models examined geological hazards, fire hazards, erosional potential, visual quality, ecological security and outdoor experience: land suitability was estimated for activities such as picnicking, camping and alpine skiing. All of the applications described above involved use of data planes derived from nonremote sensing sources, and the latter are probably of greater importance in many of the investigations. In several of the applications no remote sensing planes as such were incorporated, though they were relied upon in varying degrees to produce the individual data planes.

In using geographic information systems in the ways we have described, the estimates of land capability could be represented as contoured surfaces, or particular values could be assigned to each grid cell. In practice some form of thresholding will be used for the sake of clarity, and thus regions will be created only at this final stage.

Clearly a geographic information system may contain several planes of terrain information. Encoding and management of them separately will require time and money, and it may therefore be advantageous to combine them in some way. As an alternative it has been suggested (Dangermond 1979, ESRI 1979) that terrain data be stored in a single data plane consisting of manually created integrated terrain units, by physiographic and photomorphic methods similar to those described earlier in this chapter. Hence we see that although geographic information systems can be used in order to create regions of terrain characteristics or suitabilities, or indeed can eliminate their use altogether, there may still be an important role for regionalization and for the role of the human interpreter in its execution.

Once established, automated geographic information systems have the great advantage of flexibility. Models can be recalibrated, refined and developed, and revised estimates of land suitability can be rapidly obtained. It is possible to cope with changing administrative boundaries and the incorporation of updated data planes. Although such procedures are all technically feasible, they are achieved at no small cost for any realistically sized area. Creation of a comprehensive automated geographic information system should be undertaken only with a clear understanding of both the costs and the potential benefits.

5.5 Conclusions

Some years ago Charles Davis reviewed the land type that is the basic physiographic regional subdivision of the Michigan Land Economic Survey and concluded that such composites would in the future be of no value except for educational purposes 127

(Davis 1969). This he attributed to the growing capabilities of the computer, the growing use of statistical techniques and the development of remote sensing. In fact, regions have proved to have considerable resilience despite the rapid growth of these trends. Indeed the application of remote sensing in identifying land characteristics may be improved by physiographic regionalization. The reason for this lies in the limitations of the spectral signature model: that is the relationships between ground conditions and image characteristics change from area to area. Thus Draeger *et al.* (1973), working on crop-area statistics in the San Joaquin Valley, found that a stratified subdivision of the Landsat image by human interpreters based on gross image characteristics and a general knowledge of cropping practices, followed by a digital analysis of each stratum, was much more effective than using digital analysis alone. Two other examples of the benefits of stratification are provided by Rohde (1978) in the Red River Valley, North Dakota and near the Susitina River Alaska, both being concerned with land cover.

In the context of operational geographic information systems it might be thought that ideas of regionalization can safely be forgotten with the possibilities of retrieving data for any particular geographical location. It is certainly true that physiographic regionalization is a far less important tool than it used to be in collecting terrain data, analyzing it, and in forecasting the future potential of terrain. However, we should recall that a significant proportion of the data planes (such as pedological or geological ones) in geographic information systems are themselves composed of regions, derived in part by physiographic principles. Also terrain unit data itself may form a data plane in its own right within a geographic information system.

If we wish, we can create regions using computer-based algorithms. However, the ready availability of automated techniques of regionalization itself does not mean they will always be adopted. Visual regionalization can prove to be much more cost-effective for many tasks than computer-implemented methods. The amount of data in a single Landsat frame is enormous and consequently the computational time for even large computer systems can be very high. If small-scale map production is required, then pixel by pixel classification may be unnecessarily detailed. The power of the human interpreter with virtually random access to any part of an image, an ability to examine complex spatial image structures accompanied by access to a variety of sophisticated analogues of terrain in order to aid analysis, will all combine in the forseeable future to ensure the retention of visual physiographic methods for many applications. On the other hand, the need to merge data sets into geographically compatible forms, continuously to update temporally variable data sets and to implement increasingly complex data manipulations, will inevitably encourage the growth of computer-assisted methods.

References

Aitchison, G. D. and K. Grant 1968. Terrain evaluation for engineering purposes. In *Land evaluation*, G. A. Stewart (ed.), 125–46. Melbourne: Macmillan.

American Society of Photogrammetry 1960. *Manual of photographic interpretation*. Washington, DC: American Society of Photogrammetry.

Beckett, P. H. T. 1968. Method and scale of land resource surveys in relation to precision and cost. In *Land evaluation*, G. A. Stewart (ed.) 53–63. Melbourne: Macmillan.

Beeman, L. E. 1978. Computer-assisted resource management. *Proc. Natl Workshop on Integrated Inventories of Renewable Natural Resources, Tucson, Arizona*, 375–81.

Benn, B. O. and W. E. Grabau 1968. Terrain evaluation as a function of user requirements. In *References* *Land evaluation*, G. A. Stewart (ed.), 64–76. Melbourne: Macmillan.

Bourne, R. 1931. *Regional survey and its relation to stock-taking of the agricultural resources of the British Empire*. Oxford Forestry Memoirs no. 13.

Bowman, I. 1914. *Forest physiography, physiography of the United States and principal soils in relation to forestry*. New York: John Wiley.

Brink, A. B. A., J. A. Mabbutt, R. Webster and P. H. T. Beckett 1966. *Report of the working group on land classification and data storage*. MEXE Rep. no. 940, Christchurch, England.

Brueck, D. A. 1976. FMC's use of satellite imagery in agro-industrial development. In *Earth Resources Management, 1976 Symp., Houston, Texas*, G 320–5737, 199–245, New York: IBM.

Buttery, R. F. 1978. Modified ecoclass – a forest service method for classifying ecosystems. *Proc. Natl Workshop on Integrated Inventories of Renewable Natural Resources, Tucson, Arizona*, 157–68.

Canada Land Inventory 1970. *Objectives, scope and organisation*. Canada Land Inventory Rep. no. 1. Ottawa, Ontario: Queens Printer.

Chikishev, A. G. 1973. *Landscape indicators*. New York: Consultants Bureau.

Christian, C. S. and G. A. Stewart 1953. *General report on survey of Katherine-Darwin region, 1946*. CSIRO, Australian Land Res. Ser. 1.

Christian, C. S. and G. A. Stewart 1968. Methodology of integrated surveys. *Proc. Conf. on Aerial Surveys and Integrated Studies, Toulouse*, 233–80. Unesco.

Cooke, R. U. and J. C. Doornkamp 1974. *Geomorphology in environmental management*. Oxford: Oxford University Press.

Cowan, D. J., J. N. Bayne and D. A. Fairey 1976. *Development and applications of the South Carolina computerized land use information system*. South Carolina Land Resources Conservation Commission.

Cross, R. H. 1979. A case-study of computer-aided transmission line siting. In *Computer mapping in natural resources and the environment*, 13–34. Vol. 4 of *Harvard Library of computer graphics/1979 mapping collection*. Laboratory for Computer Graphics and Spatial Analysis, Harvard University.

Dangermond, J. 1979. A case study of the Zulia regional planning study, describing work completed. In *Urban, regional and state applications*, 35–62. Vol. 3 of *Harvard Library of computer graphics/1979 mapping collection*. Laboratory for Computer Graphics and Spatial Analysis, Harvard University.

Davis, C. M. 1969. *A study of the landtype*. Michigan: Department of Geography, University of Michigan.

Draeger, W. C., J. D. Nichols, A. S. Benson, D. G. Larrabee, W. A. Senkus and C. M. Hay 1973. Regional agricultural surveys using ERTS–1 data. In *3rd Earth Resources Technol. Satellite–1 Symp.* **1A**, 117–26.

ESRI (Environmental Systems Research Institute) 1979. *San Bernadino Forest wildland recreation study*. Redlands, California: ESRI.

Fenneman, N. M. 1916. Physiographic divisions of the United States. *Ann. Assn Am. Geogrs* **6**, 19–98.

Fridland, V. M. 1976. *Pattern of the soil cover*. Academy of Sciences of USSR, transl N. Kaner, Keter, Jerusalem.

Fu, K.- S. 1976. Pattern recognition in remote sensing on the Earth's resources. *IEEE Trans. Geosci. Electron.* **GE–14**, 10–18.

Gimbarzevsky, P. 1974. ERTS–1 imagery in biophysical studies. *Proc. 2nd Canad. Symp. on Remote Sensing, Guelph, Ontario*, 392–403.

Gimbarzevsky, P. 1978. Land classification as a base for integrated inventories of renewable resources. *Proc. Natl Workshop on Integrated Inventories of Renewable Natural Resources, Tucson, Arizona*, 169–77.

Gold, C. M. 1978. The practical generation and use of geographic triangular element data structures. In *1st Int. Advanced Study Symp. on Topological Data Structures for Geographic Information Systems*, G. Dutton (ed.). Laboratory for Computer Graphics and Spatial Analysis, Harvard University.

129

Gonzalez, R. C. and M. G. Thomason 1978. *Syntactic pattern recognition.* Reading, Mass.: Addison-Wesley.

Gonzalez, R. C. and P. Wintz 1977. *Digital image processing.* Reading, Mass.: Addison-Wesley.

Grant, K. and G. D. Aitchison 1965. *An engineering assessment of the Tipperary area, Northern Territory, Australia.* Australia: Soil Mechanics Sections, CSIRO.

Gregory, K. J. and D. E. Walling 1973. *Drainage basin form and process.* London: Edward Arnold.

Grigg, D. 1967. Regions, models and classes. In *Models in geography*, R. J. Chorley and P. Haggett (eds), 461–509. London: Methuen.

Haggett, P. 1972. *Geography: a modern synthesis.* New York: Harper & Row.

Haggett, P. and R. J. Chorley 1969. *Network analysis in geography.* London: Edward Arnold.

Harvard University 1979. *Harvard Library of computer graphics/1979 mapping collection*, 6 vols. Laboratory for Computer Graphics and Spatial Analysis, Harvard University.

Herbertson, A. J. 1905. The major natural regions: an essay in systematic geography. *Geogl J.* **20**, 300–12.

Hicks, J. P. 1977. *Managing natural resource data: Minnesota land management information system.* Lexington, Kentucky: Council of State Governments.

Hobson, R. D. 1967. Fortran IV programs to determine surface roughness in topography for the CDC 3400 computer. *Computer Contribution* **14**, University of Kansas, Lawrence.

Hobson, R. D. 1972. Surface roughness in topography: quantitative approach. In *Spatial analysis in geomorphology.* R. J. Chorley (ed.), 221–45. London: Methuen.

Isacenko, A. G. 1965. *Osnovy landsaftovenija.* Moscow: Fisiko-geograficesko rajonirovanije.

Joerg, W. L. G. 1914. Natural regions of northern America. *Ann. Assn Am. Geogrs* **4**, 55–83.

Johnson, R. J. 1968. Choice in classification; the subjectivity of objective methods. *Ann. Assn Am. Geogrs* **58**, 575–89.

Kettig, R. L. and D. A. Landgrebe 1976. Classification of multispectral image data by extraction and classification of homogeneous objects. *IEEE Trans. Geosci. Electron.* **GE–14**, 19–26.

Knapp, E. M. 1978. *Landsat and ancillary data inputs to an automated geographic information system: applications for urbanized area delineation.* CSC/TR–78/6019, Goddard Space Flight Center, Greenbelt, Md: Computer Sciences Corporation.

Knapp, E. M. and D. Rider 1979. Automated geographic information systems and Landsat data: a survey. In *Computer mapping in natural resources and the environment*, 57–68. Vol 4 of *Harvard Library of computer graphics/1979 mapping collection.* Laboratory for Computer Graphics and Spatial Analysis, Harvard University.

Kruck, W. and W. Kantor 1975. Hydrogeological investigations in the pampas of Argentina. *Proc. NASA Earth Resources Survey Symp., Houston, Texas, June 1975*, NASA TMX–58168, 2183–97.

Lankford, P. M. 1969. Regionalization: theory and alternative algorithms. *Geogl Anal.* **1**, 196–212.

Leuder, D. R. 1959. *Aerial photographic interpretation.* New York: McGraw-Hill.

Lewis, A. J. 1971. *Geomorphic evaluation of radar imagery of southeastern Panama and northwestern Colombia.* CRES Tech. Rep. 133–18, University of Kansas, Lawrence.

Linton, D. L. 1951. The delimitation of morphological regions. *London essays in geography*, 199–217. London: Longman.

Lo, C. P. 1976. *Geographical application of aerial photography.* Newton Abbot: David & Charles.

MacPhail, D. D. 1971. Photomorphic mapping in Chile. *Photogramm. Engng* **37**, 1139–48.

Mather, P. 1972. Areal classification in geomorphology. In *Spatial analysis in geomorphology*, R. J. Chorley (ed.), 305–22. London: Methuen.

MEXE (Military Engineering Experimental Establishment) 1965. *The classification of terrain intelligence.* Reports of the Combined Pool (AER), 1960–64, MEXE Rep. no. 915.

Milne, G. 1935. Some suggested units of classification and mapping, particularly for East African soils. *Soil Res.* **4**, 183–98.

Miller, L. D. and R. L. Pearson 1971. Areal mapping of the I BP grassland biome. Remote sensing of the productivity of the shortgrass prairie as input into biosystem models. *Proc. 7th Int. Symp. on Remote Sensing of Environment, Ann Arbor, Michigan*, 165–205.

Missillati, A., A. E. Prelat and R. J. P. Lyon 1979. Simultaneous use of geological, geophysical and Landsat digital data in uranium exploration. *Remote Sensing of Environment* **8**, 189–210.

Mitchell, C. W. 1973. *Terrain evaluation*, London: Longman.

Monkhouse, F. J. and H. R. Wilkinson 1963. *Maps and diagrams*. London: Methuen.

Moss, R. P. 1969. The appraisal of land resources in tropical Africa. *Pacific Viewpoint* **10**, 18–27.

Moss, R. P. 1978. *Concept and theory in land evaluation for rural land use planning*. University of Birmingham, Dept of Geography, Occ. Publ. no. 6.

Nunnally, N. R. 1969. Integrated landscape analysis with radar imagery. *Remote Sensing of Environment* **1**, 1–6.

Parry, J. T., J. A. Heigenbottom and W. R. Cowan 1968. Terrain analysis in mobility studies for military vehicles. In *Land evaluation*, G. A. Stewart (ed.), 160–70. Melbourne: Macmillan.

Peplies, R. W. and H. F. Keuper 1975. Regional analysis. In *Manual of remote sensing*, R. G. Reeves (ed.), 1947–98. Virginia: American Society of Photogrammetry.

Pocock, D. C. D. and D. Wishart 1969. Methods of deriving multifactor uniform regions. *Trans. Inst. Br. Geogrs* **47**, 73–98.

Preston, F. W. 1966. Two-dimensional power spectra for classification of land forms. In *Computer applications in the Earth sciences*, D. F. Merriam (ed.) 64–9. *Computer Contributions* **7**, University of Kansas, Lawrence.

Rayner, J. N. 1972. The application of harmonic and spectral analysis to the study of terrain. In *Spatial analysis of geomorphology*, R. J. Chorley (ed.), 283–302. London: Methuen.

Rohde, W. G. 1978. Improving land cover classification by image stratification. *Proc. 12th Int. Symp. on Remote Sensing of Environment, Ann Arbor, Michigan*, 729–42.

Savigear, R. A. G., J. R. G. Townshend, C. O. Justice, C. W. Mitchell and A. J. Parsons 1975. *Anglo-Argentinian geoscopy experiment rep.*, Section 8; *The application of Skylark space rocket photography to Earth Resources Survey*, vols 1–3. Dept. of Geography, University of Reading, UK.

Schlesinger, J., B. Ripple and T. Loveland 1978. The integration of Landsat with other natural resource data for 208 water quality planning units in south Dakota. *Proc. Amer. Soc. Photogram., Albuquerque, New Mexico*, 440–56.

Schlesinger, J., B. Ripple and T. R. Loveland 1979. Land capability studies of the south Dakota automated geographic information system (AGIS). In *Computer mapping in natural resources and the environment*, 105–14. Vol. 4 of *Harvard Library of computer graphics/1979 mapping collection*. Laboratory for Computer Graphics and Spatial Analysis, Harvard University.

Solntsev, N. A. 1962. Basic problems in Soviet landscape science. *Sov. Geog., Rev. Transl.* **3**, 3–15.

Speight, J. G. 1968. Parametric description of land form. In *Land evaluation*, G. A. Stewart (ed.), 239–50. Melbourne: Macmillan.

Spence, N. A. and P. J. Taylor 1971. Quantitative methods in regional taxonomy. *Prog. Geog.* **2**, 1–64.

Stone, R. O. and J. Dugundji 1965. A study of micro relief – its mapping, classification and quantification by means of a Fourier analysis. *Eng. Geol.* **1**, 89–187.

Story, R., G. A. Yapp and A. T. Dunn 1976. Landsat patterns considered in relation to resource surveys. *Remote Sensing of Environment* **4**, 281–303.

Strahler, A. N. 1964. Quantitative geomorphology of drainage basins and channel networks. In *Handbook of applied hydrology*, V. T. Chow (ed.), Section 4–11. New York: McGraw-Hill.

Svatos, V. 1979. An analysis of the capability of underdeveloped lands to contribute to water quality degradation. In *Computer mapping in natural resources and the environment*, 115–34. Vol. 4 of *Harvard Library of computer graphics/1979 mapping collection*.

131

Laboratory for Computer Graphics and Spatial Analysis, Harvard University.

Tansley, A. G. 1935. The use and misuse of vegetational concepts and terms. *Ecology* **16**, 284–307.

Tom, C, L. D. Miller and J. W. Christenson 1978. *Spatial land-use inventory, modelling and projection/ Denver Metropolitan Area, with inputs from existing maps, airphotos. and Landsat imagery*. NASA Tech. Mem. 7971D, Goddard Space Flight Center, Greenbelt, Md, USA.

Tou, J. T. and R. C. Gonzalez 1974. *Pattern recognition principles*. Reading, Mass.: Addison-Wesley.

Troll, C. 1971. Landscape ecology (geoecology) and biogeocenology – a terminological study. *Geoforum* **3**, 43–6.

Unstead, J. F. 1933. A system of regional geography. *Geography* **18**, 175–87.

USAEWS (United States Army Engineer Waterways Experiment Station) 1962. *A technique for mapping terrain microgeometry*. Tech. Rep. no. 3–612.

USAEWS 1965. *Terrain analysis by electromagnetic means*. Tech. Rep. no. 3–693.

USAEWS 1968. *Mobility environmental research study, quantitative method for describing terrain for ground mobility*. Tech. Rep. no. 3–726.

Vakilian, M. and P. J. Mahler 1974. A method of land evaluation in Iran. In *Approaches to land classification, Soils Bull*. no. 22, 71–6. FAO.

Veatch, J. O. 1933. *Agricultural land classification and land types of Michigan*. Michigan Agric. Exp. St. Spec. Bull. no. 231.

Viktorov, S. V. 1973. Morphometrical basis for selecting typical air photographs in geographic-indicator investigations. In *Landscape indicators*, A. G. Chikishev (ed.) 24–8. New York: Consultants Bureau.

Vink, A. P. A. 1975. *Land use in advancing agriculture*. New York: Springer-Verlag.

Vinogradov, B. V., K. I. Greenchuk, A. G. Isachenko, K. G. Raman and Yun Tsel Chuk 1962. Basic principles of landscape mapping. *Sov. Geog*. **3**, 15–20.

Williams, P. W. 1972. The analysis of spatial characteristics of karst terrains. In *Spatial analysis in geomorphology*, R. J. Chorley (ed.), 135–63. London: Methuen.

Wooldridge, S. W. 1932. The cycle of erosion and the representation of relief. *Scottish Geog. Mag*. **48**, 30–6.

Wright, R. L. 1972. Principles in a geomorphological approach to land classification. *Z. Geomorph, NF* **16**, 351–73.

132

6 The use of Landsat data for land cover inventories of Mediterranean lands

Christopher O. Justice* and John R. G. Townshend

6.1 Introduction

This chapter outlines the major methodological components of land cover mapping projects of Mediterranean lands using Landsat data, drawing in particular on work by the authors in southern Italy. The need for such land cover surveys arises from the particular problems of the Mediterranean region.

The Mediterranean lands have a wide variety of topography, lithology, soils and microclimate, resulting in a range of ecosystems from 'sub-alpine' to 'true-Mediterranean' (Unesco-FAO 1963, Unesco 1977). The intensive settlement of the Mediterranean lands since early history and their use for agriculture and forestry has led to enormous impacts on the natural vegetation (Naveh 1971); apart from arable and urban areas the land cover types are almost entirely man-induced degraded forms (Aschmann 1973). Many of the remaining areas of natural or quasi-natural vegetation, such as the Mediterranean forests and shrub lands, are highly vulnerable to destruction (Quezell 1977, Morandini 1977, Tomaselli 1977). Soil erosion occurs throughout the Mediterranean lands and woodland clearance, overgrazing and poor arable land management have considerably accelerated natural erosion rates (Paskoff 1973), though climatic changes may form the principal underlying control (Vita-Finzi 1969). There is a wide range in the quality and productivity of the agricultural land and farming methods vary considerably in sophistication (OECD 1969). Areas of intensive mechanized farming with good irrigation contrast strongly with nearby areas of subsistence farming where hand-sowing and the use of animals for ploughing are still common practice.

The socio-economic problems of the Mediterranean lands are complex and vary from country to country (Scargill 1974, OECD 1976), but it is possible to distinguish some general needs for resource inventories. The Mediterranean lands of Western Europe are less developed than their northern counterparts and efforts now are being made to reduce inequalities, for example, by improving communications, introducing industry (OECD 1972) and by modernizing agriculture and forestry through extension of irrigated areas, altering land tenure systems, and increasing mechanization. The demand for the recreational use of land is increasing throughout the Mediterranean and tourism is rapidly encroaching into the more remote regions, where the remaining areas of natural vegetation are found. These areas often include unique examples of vegetation species and ecosystems (Quezell 1977) and archaeological sites. With the invasion of these areas there is an increase

*C. O. Justice is currently a National Research Council Resident Research Associate at NASA/GSFC, Greenbelt, Maryland.

133

in forest-fire risk, which is still a primary concern of Mediterranean foresters (Morandini 1977). Unless land use changes are monitored and carefully controlled, irreparable damage will be done to the ecology and natural resources of these areas.

In several Mediterranean countries development and research programmes are currently being undertaken to define and solve some of the immediate regional problems, for example, by the Cassa per il Mezzogiorno in southern Italy (Mountjoy 1974), Ecothèque Méditerranéene (Long 1977), and the Unesco 'Man and the biosphere' project (Unesco 1977). But without accurate and up-to-date inventories of existing natural resources and careful monitoring of their use, these programmes will at best have limited success. Although these national groups are aware of the urgent need for systematic regional land evaluation and monitoring (e.g. Giacomini 1977, Cassinis 1978, Galli de Paratesi 1977) insufficient is being done to examine the applicability of new data collection methods or to introduce these methods to the relevant regional bodies.

6.2 Use of remote sensing in Mediterranean lands

Existing methods for collecting land resources data largely consist of ground collection at a local level by foresters, engineers and agricultural extension workers, and it is necessary to incorporate these data into a regional data collection system if valid planning decisions are to be made at a regional level. Ideally in the future a series of geographic information systems should be established for each country to integrate both local and regional data, but at present it is at least possible to improve and consolidate the present data collection systems by adopting some of the newly available remote sensing techniques. For certain countries, vegetation and land use maps were produced at a regional level in the 1960s using aerial photographs, e.g. Italy, Morocco, France, and Spain (Winch 1976). Many of the maps that were produced are now out of date and because of the broad classes chosen are largely unsuitable for present planning purposes.

The cost and time taken in acquisition make the use of conventional aerial photography for base-line regional surveys and yearly land use monitoring impractical for many purposes. The application of newer remote sensing methods to resource surveys is outlined in Chapter 2 and although many remote sensing techniques are still experimental, sufficient understanding and expertise has been acquired for these to supplement existing data collection systems (Reeves 1975, Sabins 1978, Nunnally 1974). Satellite imagery of Mediterranean areas has been obtained since 1972 through both the NASA Landsat (NASA 1976) and Skylab (NASA 1975) programmes (Ch. 2). Both Skylab S190 multispectral photography and Earth Terrain Camera (ETC) photography were taken but due to the sporadic coverage of the data, few applications experiments have been undertaken (e.g. Bodechtel *et al.* 1975, Cassinis *et al.* 1975). The Skylab Earth Terrain Camera provided the first high resolution (*c.* 21 m) Earth resource satellite imagery of Mediterranean areas (Fig. 6.1). The few high resolution colour and colour infrared Skylab images of Mediterranean areas provided users with an indication of the fine ground detail that it is possible to record from space altitudes. Although having coarser resolution, the Landsat multispectral scanner system's (MSS) coverage of the Mediterranean has been complete and repetitive. Landsat data have been shown to provide an effective way of collecting environmental information at a regional level in many environments (van Genderen & Lock 1977b, Guernsey & Mausel 1978, Miller & Williams 1978, Miller *et al.* 1978, Tom *et al.* 1978, Peplies & Keuper 1975,

Figure 6.1 Photograph from the Skylab Earth Terrain Camera of the area around the junctions of the Verdon and Asse rivers with the Durance in southwest France.

Williams & Carter 1976, NAS 1977) but relatively limited applications of these data have been made with respect to land cover mapping in Mediterranean environments. Most of the early evaluations of Landsat data for application to Mediterranean land use were undertaken by visual image interpretation methods as shown in studies in Greece (Yassoglou *et al.* 1973), in southern France (Caballe *et al.* 1974, Rey 1974), in Italy (Justice *et al.* 1976, Bodechtel *et al.* 1975) and in Spain (Fernandez 1977). Analysis by photo-interpretative methods is sufficient to delimit major land cover types at relatively small scales of 1:5000000 and less, though attempts using such methods have been made to delimit more specific categories such as citrus trees and rice fields in eastern Spain by De Sagredo and Salinas (1973). Table 6.1 is taken from Bodechtel *et al.* (1975) and shows levels I–IV land use categories as defined by Anderson *et al.* (1972), which can be 'distinguished' in Italy by photo-interpretation of Landsat imagery.

The development of computer-assisted methods for the analysis of Landsat data has led to an increase in both quantitative and repeatable results. Digital Landsat tapes contain potentially far more data than are available from the photographic products and enable the application of automated image enhancement and classification procedures. Various approaches to computer-assisted land cover mapping have been adopted, using supervised classification (Dejace *et al.* 1977, Lenco 1978, Le Toan *et al.* 1977, Justice 1978b), unsupervised classification (Allan 1975, Fontanel *et al.* 1975, Labrandero 1976, Le Toan & Megier 1978) and hybrid classification methodologies (Townshend & Justice 1980). Difficulties have been found in some Mediterranean studies in obtaining sufficiently high accuracy of classification, because of the complexity of the land cover types as reported by Cassinis *et al.* (1975) working in volcanic terrain near Naples, and by Allan (1975) in central Spain. Nevertheless, some considerable success has been achieved as is shown by the Agreste project in northern Italy and southern France (CEC 1978). Using multitemporal Landsat imagery of southern France to discriminate between

135

Table 6.1 Land cover categories distinguishable on Landsat 1 imagery of Italy by photo-interpretative methods (Bodechtel *et al.* 1975) using the standard levels devised by Anderson *et al.* (1972).

Level I	Level II	Level III	Level IV
Agricultural land	arable	grain crops	wheat rice
	plantations	truck crop vineyards orchards olive groves citrus nut plantations	
Woodland	deciduous coniferous		chestnuts Mediterranean oak macchia spruce pines stone pines larches
Water	lakes	natural artificial	shallow deep
	rivers		high *vs* low suspen-ded sediment
Barren land	bare rock	volcanic sedimentary	limestone sandstone marls shales

rice, wheat and vineyards, Le Toan and Megier (1978) obtained a 90% correct classification. In northern Italy using Landsat data taken between sowing and emergence and extrapolating from a large test area, the same authors identified rice fields with an accuracy 'not inferior to 95%'. Dejace (1978) using single date Landsat imagery identified beech trees in northern Italy with a 63% classification accuracy; this accuracy increased to 83% when spectral channels were ratioed. CEC (1976) reported a 70% correct classification for poplar forests in northern Italy having a ground coverage of more than 25%.

Lenco (1978) undertook a supervised classification of a single Landsat scene of an area in south-west France, with an overall classification of 94% for fifty land use categories. By regrouping the classes, he achieved 100% correct classification for agricultural land, 92% for forest, 98% for evergreen, 92% for deciduous, 83% for mixed forest, and 89% for garrigue vegetation.

On the basis of such work both in Mediterranean areas and elsewhere, the stages which need to be executed for successful application of Landsat data for land cover inventories are now outlined.

6.3 Procedures for using Landsat digital data in the Mediterranean region

Many of the procedures developed elsewhere in the world can be used in the Mediterranean region, but local conditions dictate that they must be applied with

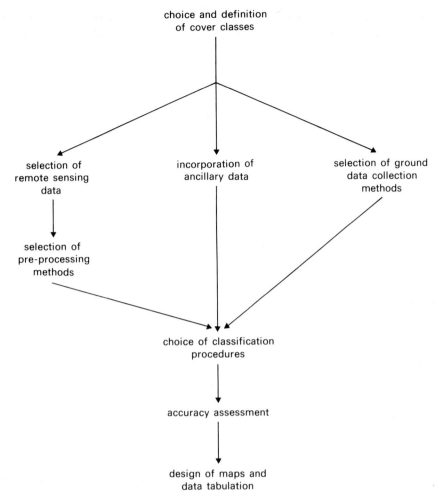

care and modified appropriately. In particular the small size of homogeneous areas of single land cover types in relation to the current resolving power of satellite sensors provide problems of identification and location of pixels and corresponding ground data sites. Moreover, many areas are not characterized by simple land cover types, nor frequently are the types separated by abrupt boundaries but merge one with another. Of course such conditions are also characteristic of areas outside the Mediterranean, but they are less typical of areas where Landsat data were originally and have been most commonly applied. We wish to demonstrate both from work carried out in Mediterranean lands and elsewhere how these difficulties can in part be alleviated and how digital data in particular can be used. The use of digital data rather than the photographic products is stressed since the former offer prospects for mapping at far larger scales and with higher accuracy.

The successful execution of any data collection system for land cover description demands that a sequence of decisions and procedures are made as described in outline in Figure 6.2.

6.3.1 *Choice of cover classes*

Choice of the specific cover classes to be designated is largely dependent on the ultimate use of the information, but certain principles underlie their selection. In particular the trade-off between precision of definition and accuracy of discrimination is important. For example, woodland and non-woodland classes can be discriminated with a higher accuracy than can woodland types, though the discrimination is by definition less precise. The degree of refinement of the classes that is feasible will be dependent on the potential discriminating ability of the Landsat scanners, and on the sophistication of the classification techniques. It may be that the categories which can be discriminated using Landsat data are less precise than those ideally required and potential users of Landsat data should be aware both of the type of categories obtainable from such data and the way in which they can be related to more detailed land cover classes derived from other data sources. If a hierarchical scheme of classes is used the latter should be facilitated. The most common scheme is that derived by Anderson *et al.* (1972) for land use mapping of the US Geological Survey, but in several respects it is poorly suited to many Mediterranean environments and should be used with care. In particular, where several cover types exist at an individual site, as often happens in the Mediterranean region, classification into a single cover type, such as the dominant one, may be a gross oversimplification. Classes can be defined to give an indication of the dominant and secondary cover types at a site, with names such as 'mature deciduous oak woodland with evergreen shrub understorey', but it is preferable to describe the percentage of cover types at each site (Justice 1978b).

6.3.2 *Role of ancillary data*

Apart from ground data, other forms of data are usually required, even where remote sensing provides the principal source of information. These may comprise no more than existing topographic maps or aerial photographs to aid the location of training sites (Section 4.9). Alternatively, ancillary data planes may be incorporated to provide extra dimensions in the feature space in order to aid discrimination of cover types (Tom *et al.* 1978). Prior to a study, the extent and reliability of the existing data base should be examined. Study of sources such as topographic, geological, pedological and vegetational maps will help in understanding the spatial relationships of cover types with other terrain properties and hence determine whether it is worth incorporating such ancillary data planes to supplement the multispectral data in land cover discrimination. Such preliminary examinations are important because the time and cost involved in producing data planes compatible with spectral data can be high. In their studies of the applicability of Landsat data in southern Italy, the authors noted that several species had distinct habitats related to

Table 6.2 Statistical relationships between terrain properties using Kendall's tau for 180 sites in Basilicata, southern Italy (Justice 1978a).

Height above sea level

Lithology	*Slope angle*	*% tree cover*	*% bare soil*	*% herb. vegn*	*% deciduous vegn*
0·2295	0·2174	0·2392	−0·1733	−0·1609	0·1734
0·012	0·001	0·001	0·001	0·002	0·001

Slope angle

Lithology	*% tree cover*	*% shrub vegn*	*% herbaceous vegn*	*% deciduous vegn*
0·3023	0·2696	0·2110	−0·2269	−0·1430
0·001	0·001	0·001	0·001	0·005

Key 0·2295 = Kendall's tau.
0·001 = Two-tailed significance.

broad classes of lithology and soil moisture and found statistically significant relationships between selected terrain variables and cover type (Table 6.2). Flouzat (1978) demonstrated the relationships between the distribution of fir forests in southern France, and slope angle, aspect and elevation and suggested that such *a priori* data can be used to aid classification of Landsat imagery (Fig. 6.3). Altitudinal control of vegetation types is common throughout the Mediterranean region and hence makes elevation a good potential discriminant of cover types, but it should be noted that the altitudinal stratification of woodland types varies with latitude (Morandini 1977) and aspect (Sadowski & Malila 1977).

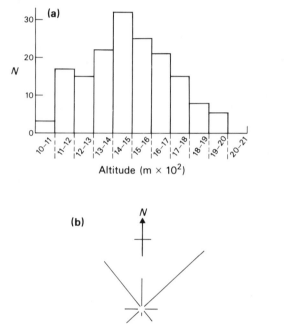

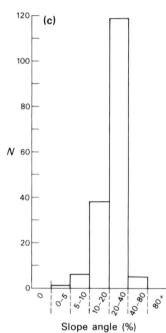

Figure 6.3 Distribution of fir forests in relation to (a) altitude, (b) aspect and (c) slope angle (from Flouzat 1978).

6.3.3 *Selection of images*

The Landsat ground receiving station at Fucino, Italy obtains data for the whole of the Mediterranean region (Fig. 6.4). Since the opening of this station, data are collected for the large majority of Landsat passes, so that the user will usually be able to select images from several dates, though cloud cover and image quality will have to be considered before ordering or carrying out an extensive ground data collection programme.

It is important to know the seasonal relationships between the cover types being studied, so that the optimum date for discrimination can be obtained. Table 6.3 shows the Landsat images available for the CEC (1978) rice study in northern Italy and the associated ground conditions. The optimum time for interpretation was found to be the three or four week period between sowing and emergence of the rice crop. The optimum data for discrimination will of course not be the same for all cover types; for example, a winter scene will provide good discrimination between evergreen and deciduous vegetation. One way to establish the date best suited to a particular study is to produce a vegetation or crop calendar showing the periods of sowing and harvesting and maximum foliation (e.g. Table 6.4). The Landsat spectral response to the seasonal variations should be considered carefully when

139

Figure 6.4 Coverage of the Landsat ground receiving station, Fucino, Italy.

Table 6.3 Development stages of rice plants (for varieties *Balilla* and *Balilla GG*) and soil conditions in relation to the date of Landsat passes yielding suitable imagery (CEC 1978).

Date of Landsat pass	Development stage	Soil conditions
22 April	Sowing (17–18 April)	Under water
10 May	No information collected	Under water
28 May	Emergence	Under water
15 June	Beginning of tillering	Under water
3 July	Tillering – field coverage not complete	Under water
21 July	Tillering – complete field coverage – beginning of booting	Under water
8 Aug.	Late flowering (beginning 5–6 Aug.)	Under water
26 Aug.	Late milk stage	Drained
13 Sept.	Late dough stage	Drained
1 Oct.	Harvest (29 Sept. – 6 Oct.)	Drained
19 Oct.	Harvested	Drained

Table 6.4 A generalized crop calendar for part of the Agri Valley, Basilicata, southern Italy (Justice 1978a).

Crop \ Month	Jan.	Feb.	Mar.	Apr.	May	June	July	Aug.	Sept.	Oct.	Nov.	Dec.
Wheat					Stress	Harvest			Plough		Sow	
Maize				Sow				Harvest				
Vetch			Harvest								Sow	
Vines						Irrigation	Stress		Harvest			
Olive			Cutting				Stress					
Alfalfa						Stress	Harvest			Sow		

selecting the data for the analysis. D'Aubarede and Avizanda (1976) examined the reflectance curves for pine species in central Spain and showed that it was not possible to differentiate between *Pinus pinea* and *Pinus pinaster* using Landsat data, as their flowering periods occurred at the same period. However, by careful image selection it would be possible to differentiate between *Pinus sylvestris* and *Pinus pinea* as their flowering periods occur two months apart. Studies in many parts of the world have shown that selective use of multitemporal data can improve classification accuracies (e.g. Anuta & Bauer 1973, Williams & Haver 1976, Lenco *et al.* 1978, Flouzat & Megier 1978) but the computing time in terms of image processing and registration is considerably higher. Table 6.5 shows the improved classification results obtained by using multitemporal data for discrimination of rice, wheat and vines (Le Toan & Megier 1978).

Care should be taken to avoid scenes with low Sun angles since the resultant 'topographic effect' on the imagery (Holben & Justice 1979) may strongly reduce the clarity with which land cover is displayed. Such effects are often very marked in rugged Mediterranean terrain. A visual impression of the significance of Sun angle can be seen in Figure 6.5 and a quantitative estimate of the interaction between slope angle, aspect and Sun angle is given in Figure 6.6 for the two dates with the maximum difference in Sun elevation.

Table 6.5 The optimum combinations of spectral bands and dates for discrimination of rice, wheat and vines in S. France (Le Toan & Megier 1978).

	Computed (numerical values indicate number of pixels)								
Actual	Rice	Wheat	Vines	Rice	Wheat	Vines	Rice	Wheat	Vines
Rice	209	11	12	214	5	13	214	5	13
Wheat	3	172	33	12	176	20	7	183	18
Vines	12	7	79	3	9	86	3	6	89
	% accuracy 85·50			% accuracy 88·48			% accuracy 90·33		
Spectral band	MSS 7, MSS 5, (11 Aug)			MSS 7, MSS 5, MSS 4 (11 Aug) MSS 7 (6 July)			MSS 7, MSS 5, MSS 4 (11 Aug) MSS 7 (6 July) MSS 5 (23 July)		

(a)

(b)

Figure 6.5 Landsat images of part of Campania and Basilicata, southern Italy. (a) 8 August 1972, Sun elevation 55°, azimuth 120°; (b) 6 November 1972, Sun elevation 30°, azimuth 154°: both channel 5 images. Area depicted: 145 × 135 km. Clearly Sun elevation can have a very marked effect on the appearance of satellite imagery and its relative utility for different tasks. Whereas image (a) is to be preferred for land cover mapping, image (b) has much greater value for geomorphological or geological mapping. NASA images E-1016-09113-5 and E-1106-09121-5.

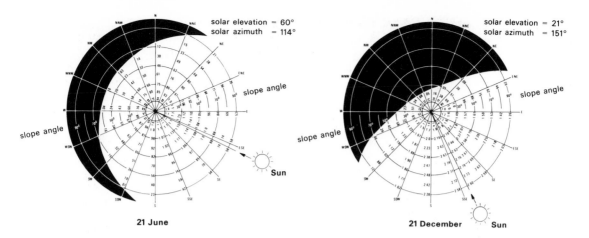

6.3.4 *Ground data collection*

Before collecting the ground data themselves, the sampling schemes must be designed to produce a representative set of training and testing sites (Ch. 3). Where there is a diversity of terrain, the problem of representing the surface conditions is compounded. One way of overcoming this is to produce a preliminary stratification of the area on the basis of physiographic cover units and then undertake sampling within each unit (Ch. 5 and Rohde 1978). For example, Hoffer *et al.* (1978) in a study in Colorado, USA, found that survey units of the US Forestry Service provided a useful stratification of the terrain into physiographic cover units for estimating forest acreage using Landsat data. If parametric statistics are to be used in the classification procedure, an independent and random sample must be obtained (Ch. 3). In southern Italy, Justice (1978b) found that a non-aligned systematic random sampling scheme was most suitable for representing the cover of widely varying areas. The location of the sampling frame was based on a preliminary terrain unit subdivision and the systematic grid size was determined by the frequency of terrain variation occurring in the study area. If one is only interested in certain specific cover types then a more purposive sampling scheme should be used. For land cover studies van Genderen and Lock (1977b) recommended a stratified random sampling scheme and point out that sample sizes should be sufficiently large to enable statistically meaningful results to be obtained. The size of individual sample sites should in part be determined by the ease with which they can be located on the Landsat data. Justice (1978b) used uniform field sample sites of 200 m square to enable satisfactory pixel location. Townshend and Justice (1980) found that larger areas of uniform cover were required for identification of cluster classes in an unsupervised classification study of the Basilicata area.

The great mixture of vegetation types and species found in several Mediterranean ecosystems and the widespread inter-culture of tree and arable crops, make site description a complex problem. Use of aerial photography as surrogate ground data without field training can lead to misinterpretation of cover conditions. For example, distinction between immature deciduous woodland and evergreen maquis is difficult using conventional panchromatic aerial photographs, even if there is prior field knowledge concerning their likely distribution. When aerial photographs are used to provide surrogate ground data, there should be a preliminary testing phase to assess the accuracy of the photo-interpretation.

Use of aerial photographs is always recommended, even if they are not directly used to provide ground data, as they provide a spatial portrayal of the surface,

Figure 6.6 Direct incoming solar radiation for different slope angles (represented by concentric circles) and aspects relative to a horizontal surface. The values represent incidence values (see p. 48) and the black areas represent slopes in shadow which are illuminated only by diffuse light.

143

Table 6.6 Types of environmental change and factors affecting the timing of data collection for interpretation of remotely sensed data in Mediterranean environments, based on a study in Basilicata Province, S. Italy.

Rate of change	Type and description of change	
Short-term changes (daily/weekly)	Agricultural	Change in crops during the growing season Time of growing, ploughing, harvesting Crop stress Drought and irrigation
Medium-term changes (monthly/seasonally)	Agricultural	Crop rotation Pruning and thinning of fruit crops Grassland seasonal variation
	Natural/Semi-natural	Evergreen/deciduous phenology Grazing of herbaceous and shrubland
	Lakes and dams	Lake level change and sedimentation
Long-term change (yearly and longer)	Agricultural	Change from arable to fruit/vines Change from pasture to arable Change from poor arable to pasture Regeneration of old olive groves
	Woodland	Annual growth (immature to mature) Deforestation (clear cut or selective) Reafforestation and regeneration
	Semi-natural	Maquis clearance and degeneration Maquis regeneration Erosion (sheet/gulley/mass movement)
	Lakes and dams	Construction and sedimentation
	Anthropomorphogenic	Highway construction, quarrying etc.
	Catastrophic	Fire Landslides Fluvial erosion and deposition

intermediate between ground and satellite information and these greatly aid ground location of pixels. Lenco (1978) in a study of the Garonne Valley, France, shows how multispectral aircraft data provide a useful aid to the interpretation of multispectrally similar satellite data.

The timing of the ground data collection relative to the Landsat pass should be dependent on how rapidly the cover types change. For example, analysis of agricultural areas subject to rapid change will require ground data collection near to the time of pass. The environmental factors likely to affect the timing of ground data collection in Mediterranean areas are listed in Table 6.6.

6.3.5 *Pre-processing of data*

Pre-processing of Landsat data should be undertaken prior to multispectral classification and should provide improved data for classification and subsequent analysis. Variations in the six Landsat sensor detectors for each channel may sometimes lead to a marked visual 'six-line' banding effect across an image, especially if ratioing is carried out. Algorithms exist which help eliminate such banding (e.g. Horn & Woodham 1978), and should be applied to the data, though it is difficult to eliminate these effects over large areas using constant correction parameters.

Geometric corrections can also be applied to the satellite data (Anuta 1973, Rifman 1973) to eliminate geometric distortions within the data and to fit the Landsat data to specific map projections. Methods for achieving this map registration are described by several workers (e.g Graham 1977, Kirby & Steiner 1977, Hardy 1978, Horn & Woodham 1979) but difficulties may occur where there is a high range of relief which increases greatly the need for ground control points.

Differences in spectral response due to the topographic effect can be reduced by the ratioing of selected multispectral channels which also reduces atmospheric effects (Vincent 1973, Schrumpf 1975, Goetz *et al.* 1975, Holben & Justice 1980). Figure 4.4 shows the effects of ratioing a Landsat image on the Gulf of Ionico, southern Italy. Note that ratioing reduces the topographic effect but increases the six-line banding effect. Differential atmospheric and Sun angle effects may present particular problems for multitemporal Landsat studies and they should preferably be removed prior to classification: Ahern *et al.* (1979) describe transformations to reduce 60–90% of such variations.

Fontanel *et al.* (1975) in a study in southern France showed how principal components transformations can reduce the data used in classification and enhance the interpretation.

6.3.6 *Classification methods*

The classification procedure adopted will be dependent on the algorithms available to the user. Essentially there are two alternative approaches, namely supervised and unsupervised (Ch. 4). However, several hybrid systems have been developed which incorporate aspects of both (Fleming *et al.* 1975 and see Table 4.4).

A supervised approach was used by Justice (1978a) on imagery from August and produced an 85% correct classification for seven major cover types in Basilicata Province, southern Italy (Table 6.7). These results were based solely on discrimination within the training sites and it is likely that lower accuracy figures would be obtained if extrapolation over the study area were undertaken. It was found generally that ratio data provided the optimum discrimination of cover types. An unsupervised classification of Landsat data of the region around Madrid, Spain in July was undertaken by Labrandero (1976). Although no accuracy statements were presented, the graph of the separability of cluster class suggests that discrimination between many of the classes would be possible (Fig. 6.7) even with only two bands.

Townshend and Justice (1980) undertook classification of part of the Basilicata study area using a hybrid method intermediate between the supervised and unsupervised approaches, namely the monocluster block unsupervised approach (see Table 4.4). This involved grouping a series of sample areas, carrying out cluster analysis on the combined samples, identifying resultant classes and then extrapolating over the whole area using the training statistics. Overall accuracies of 85% and 65% were obtained for two testing areas of varying complexity for ten surface cover classes, the accuracy being dependent on the complexity of the test areas, as described by relief ruggedness and size of cover units. The authors found that extreme care had to be taken in identifying the range of cover conditions represented by the cluster classes. This study failed to demonstrate that an unsupervised approach was demonstrably superior to a supervised approach for areas of complex terrain, as suggested by other workers (e.g. Hoffer & staff 1975, Flemming *et al.* 1975).

The ability to discriminate between cover types varies in different areas due to the high variability of terrain and wide range of ecotypes (Justice 1978a, Townshend & Justice 1980). Localized microclimatic effects, irregular crop calendars, and a wide

Table 6.7 Summary of discrimination results for training sites, Basilicata Province, s. Italy (Justice 1978a).

Number of classes	Types of classes	Classification accuracy			Optimum channel combination
		good (>85%)	moderate (70–85%)	poor (<70%)	
3 (defined by > 50% cover)	Vegetated/non-vegetated water	★			5–6/5+6
3 (defined by > 50% cover)	Trees/shrub/ herbaceous/ non-vegetated	★			7–5/7+5
2	Arable/non-arable	★			5–6/5+6
3	Tree density (<24%/25–84%/>85%)		★		7–6/7+6
2	Tree height (<5 m/>5 m)	★			5–6/5+6
4	Herbaceous with different secondary covers			★	7–5/7+5
2	Managed pasture/rough grazing	★			5–6/5+6
7	General surface cover classes	★			7–5/7+5
10	Lithological units			★	7–5/7+5

range of farming types cause considerable variations in planting and harvesting dates over relatively small areas which combined with a variety of crop conditions make discrimination of crop types and agricultural areas a difficult problem.

Justice (1978a) and Holben and Justice (1979) show that one way to improve classification accuracies is to stratify the area using selected terrain parameters. This will reduce the variation in radiance associated with a given cover type. Improvements in accuracy by stratifying within coarse elevation and slope categories are presented in Table 6.8. When site elevation was included with the sensor data as a discriminating variable in discriminant analysis of major cover types, an absolute

Figure 6.7 Classes derived by unsupervised classification of a Landsat scene of Madrid, plotted in two dimensions of the feature space. The high correlation between the channels with respect to the class centroids, indicates that there is considerable redundancy in the discriminatory powers of the two channels (from Labrandero 1976).

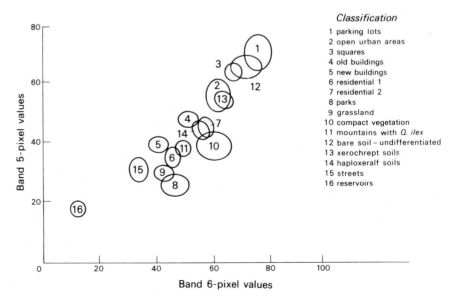

Classification
1 parking lots
2 open urban areas
3 squares
4 old buildings
5 new buildings
6 residential 1
7 residential 2
8 parks
9 grassland
10 compact vegetation
11 mountains with Q. ilex
12 bare soil – undifferentiated
13 xerochrept soils
14 haploxeralf soils
15 streets
16 reservoirs

Table 6.8 Discrimination accuracies (training set) for dominant woodland, shrubland, herbaceous and bare soil cover types within selected terrain strata, Basilicata Province, southern Italy (Justice 1978a).

No stratification		89%	
Slope strata	< 5°	5–20°	> 20°
	100%	95·8%	100%
Elevation strata	< 500 m	500–1000 m	> 1000 m
	100%	91·7%	100%

Table 6.9 Classification matrix from a monoblock cluster analysis of a test site in Basilicata Province (Townshend & Justice 1980). (Errors of omission are found by summing horizontal elements and errors of commission by summing vertical elements, excluding the principal diagonal in both cases.)

	Predicted Actual	1	2	3	4	5	% accuracy for each class
Herb. with trees and base ground	1	18	2	0	1	2	78
Herb. with trees and shrubs	2	5	18	1	3	2	62
Decid. woodland	3	5	1	108	1	1	93
Shrubs + trees with herb. (< 50%)	4	4	0	5	8	0	47
Shrubs + trees with herb. (> 50%)	5	1	1	1	10	17	56

Overall accuracy 79·8%
Combining 4 and 5 overall accuracy 84·7%

improvement of 9% was obtained (Justice 1978a). An even larger improvement of 27% was achieved in incorporating terrain data by Strahler *et al.* (1978) in southern California. There are essentially two methods of incorporating terrain data within the classification procedure, either by using prior probabilities in the discriminant analysis or by using digital terrain data as an additional data plane (Anuta 1976). The advantages and disadvantages of the two methods are described in detail by Strahler *et al.* (1978) who concluded that overall the former was preferable.

6.3.7 *Assessment of accuracy*

Accuracy assessment can only be fully undertaken after statistics from the training sites have been used to classify at least parts of the remaining area. Figures for accuracy are obtained by comparing the predicted classification of a number of 'testing sites' with the actual ground classes. This again requires accurate ground location of the pixel data for the testing sites which is often difficult as pointed out above. Berkebile *et al.* (1976) recommended location of ground sites using registered aerial photographs and line printer output. Townshend and Justice (1980) found that the accurate location of classified pixels on aerial photographs using a stereo comparator was difficult, due to relief distortion and that an intermediate stage of 147 transferring the ground classes from aerial photographs to a topographic map was required. A sufficient number of 'test sites' should ideally be used to render the accuracy assessment statistically valid for each cover type (van Genderen & Lock 1977a). Classification accuracy is usually described in terms of percentage correct classification and is usually presented in the form of errors of commission and omission (Table 6.9). When reviewing classification accuracies, care should be taken to note the number of samples tested and make sure that the results relate to classification of all the study area rather than just the training sites.

6.4 Prospects for applications of remote sensing

The future prospects for remote sensing data applications to surface cover mapping in Mediterranean environments are favourable. Continued coverage and availability of Landsat data for Mediterranean areas is ensured through the European Space Agency (ESA) and the existing archives will provide a valuable historical record for future studies of environmental change.

One of the biggest criticisms of Landsat data for applications to Mediterranean areas has been that the 79 m resolution is too coarse for the small plot sizes (Galli de Paratesi 1977). However, Landsat 3 is currently collecting return beam vidicon (RBV) data with a nominal resolution of 24 m (RCA 1977). Townshend *et al.* (1979) made a preliminary assessment of this data of Basilicata for woodland mapping and drainage network analysis. Discrimination of spectrally high contrast oak woodland was possible at over 90% accuracy. An example of the RBV imagery for southern Italy is shown in Figure 6.8. Improved spatial resolution of satellite data will have a large impact on areas of complex terrain and the proposed spectral characteristics of the Landsat D sensors will improve discrimination of vegetation types (Tucker 1978, Salomonson 1978). The proposed generation of satellites with pointing multilinear array sensor technology, such as the French SPOT (système probatoire pour observation de la terre) satellite (CNES 1977, Gaubert 1978) and the NASA MRS (multi-linear array resource sampler) system (ORI 1978, Schnetzler & Thompson 1979) will both provide 10–20 m resolution data. The French SPOT system is being designed specifically to meet European needs and will have a priority for European coverage (CNES 1978).

Although there is a slow realization that remotely sensed data can contribute to natural resource inventories in Mediterranean areas, it is necessary for the national and regional planning authorities to incorporate the new techniques more rapidly into existing data collection systems. There has been a tendency in the past to oversell the value of new remote sensing techniques and it is now the responsibility of the European remote sensing community to demonstrate the true value of remotely sensed data. Remote sensing at present certainly cannot provide the detailed information required in many ecological studies. On the other hand many of the significant changes in vegetation cover in the region will be missed by a slavish adherence to highly local investigations. Regional organizations incorporating remotely sensed data and analysis methods must carefully appraise existing Landsat studies and in some cases modify the existing methods, or run pilot projects to establish the optimum system for their own requirements. There are now a wealth of possibilities for improving and updating the existing natural resource inventories and creating new inventories where previous ones do not exist.

(a)

(b)

Figure 6.8 Comparison of (a)
a return beam vidicon image
(Landsat 3) with (b) a
multispectral scanner image
(Landsat 2) of part of Puglia
near Taranto. NASA images
E-30178-08552-B and E-2191-
08541-5. Images were obtained
in different years but at similar
calendar dates. Area depicted:
30×25 km.

References

Ahern, F. J., P. M. Teillet and D. G. Goodenough 1979. Transformation of atmospheric and solar illumination conditions on the CCRS image analysis system. *Proc. Symp. on Machine Processing of Remotely Sensed Data, Purdue University, Indiana*, 34–52.

Allan, J. A. 1975. Land use in the Merida region of Spain: an application of the LARS system in a complex agricultural area using ERTS 1. Paper presented at *2nd Annual Conf. Remote Sensing Soc., Silsoe, UK*.

Anderson, J. R., E. E. Hardy and J. T. Roach 1972. *A land use classification scheme for use with remote sensor data*. US Geol surv. circ. 671, Washington.

Anuta, P. E. 1973. *Geometric correction of ERTS–1 digital multispectral scanner data*. LARS Inf. Note 103073.

Anuta, P. E. 1976. Digital registration of topographic data and satellite MSS data for augmented spectral analysis. *Proc. 42, Am. Soc. Photogramm.*, 180–7.

Anuta, P. and M. Bauer 1973. *An analysis of temporal data for crop species classification and urban change detection*. LARS Inf. Note 110873.

Aschmann, H. 1973. Man's impact on several regions with Mediterranean climates. In *Mediterranean type ecosystems: origins and structure*, F. di Castri and H. A. Mooney (eds), 364–71. New York: Springer-Verlag.

Berkebile, J., J. Russell and B. Lobe 1976. *A forestry application of man–machine techniques for analyzing remotely sensed data*. LARS Inf. Note 012376.

Bodechtel, J. *et al.* 1975. Application of Landsat and Skylab data for land use mapping in Italy. *Proc. NASA Earth Resources Surv. Symp., Houston, Texas*, 1863–86.

Caballe, G., B. Lacaze and G. Long 1974. Étude qualitative de quelques images ERTS–1 du project Golion aspects écologiques des milieux terrestres. *Eur. Earth Resources Satellite Experiments, Frascati, Italy*. B. T. Battrick and N. T. Duc (eds), ESRO Sp. 100, 363–70.

Cassinis, R. 1978. Italian activities in the domain of remote sensing. *Proc. 12th Int. Symp. on Remote Sensing of Environment, Ann Arbor, Michigan*, 145–58.

Cassinis, R., G. M. Lechi and A. M. Tonelli 1975. Application of Skylab imagery to some geological and environmental problems in Italy. *Proc. NASA Earth Resources Surv. Symp., Houston. Texas*, **1–B**, 851–67.

CEC (Commission of European Communities) 1976. *Agricultural investigations in northern Italy and southern France*. Follow on investigation no. 28790, 2nd prog. rep.

CEC 1978. *Agreste project: agricultural resource investigations in N. Italy and S. France*. Final Report, NASA Investigation no. 28790.

CNES (Centre National d'Etudes Spatiales) 1977. *Caractéristiques principales du satellite national d'observation de la terre*. Document CST/D/863, Centre Spatial de Toulouse.

CNES 1978. *The Earth observation test system: a general description*. Centre Spatial de Toulouse.

D'Aubarede, J. and P. Avizanda 1976. Differentiation of conifer species in central Spain by digital analysis of Landsat data. In *Thematic mapping, land use, geological structure and water resources in central Spain*, N. De las Cuevas (ed.), 154–70. NASA Final Rep. E77–10167.

Dejace, J. 1978. Identification of beech forests. In CEC (1978), 137–45.

Dejace, J., J. Megier and W. Mehl 1977. Computer aided classification for remote sensing in agriculture and forestry in northern Italy. *Proc. 11th Int. Symp. Remote Sensing of Environment, Ann Arbor, Michigan*, 1269–78.

De Sagredo, F. L. and F. G. Salinas 1973. Identification of large masses of citrus fruit and rice fields in eastern Spain. *Symp. on Significant Results from ERTS–1*, **1A**, 35–6.

Fernandez, E. C. 1976. Application of Landsat 2 data to land use mapping in central Spain. In *Thematic mapping, land use, geological structure and water resources in central Spain*, N. De las Cuevas (ed.) 216–21. NASA Final Rep. E77–10167.

Flemming, M. D., J. S. Berkebile and R. M. Hoffer 1975. *Computer aided analysis of Landsat–1 MSS data*. LARS Inf. Note 072475.

Flouzat, G. 1978. Identification and inventory of fir forests with Landsat data. In CEC (1978), 146–58.

Flouzat, G. and J. Megier 1978. Identification and inventory of poplar groves. In CEC (1978), 104–45.

Fontanel, A., C. Blanchet and C. Lallemand 1975. Enhancement of Landsat imagery by combination of multispectral classification and principal component analysis. *Proc. NASA Earth Resources Symp., Houston, Texas*, **1B**, 991–1012.

Galli de Paratesi, S. 1977. Development and prospects of remote sensing applied to European renewable Earth resources. In *Earth observation systems for resource management and environmental control*, D. J. Clough (ed.), NATO Conf. ser. 11, 229–41. New York: Plenum.

Gaubert, A. M. 1978. SPOT: the French trial earth observation system. *Proc. 12th Int. Symp. on Remote Sensing of Environment, Ann Arbor, Michigan*, 289–96.

Giacomini, V. 1977. Conservation of terrestrial ecosystems in the Mediterranean region. *Nature and Resources* XIII, 2–6.

Goetz, A. F. H., F. C. Billingsley and A. R. Gillespie 1975. *Application of ERTS images and image processing to regional geologic problems and geologic mapping in N. Arizona*. NASA Tech. Rep. 32–1597, Jet Propulsion Laboratory, California, USA.

Graham, M. H. 1977. *Digital overlaying of the universal transverse Mercator grid with Landsat derived products*. NASA Tech. Mem. 58200.

Guernsey, L. and P. W. Mausel 1978. Application of remote sensing in regional planning. In *Introduction to remote sensing of the environment*, B. F. Richason Jr (ed.), 319–63. Iowa: Kendall Hunt Publishing.

Hardy, J. H. 1978. Land cover studies using MSS digital data: cartographically corrected lineprinter output. *Proc. 12th Int. Symp. on Remote Sensing of Environment, Ann Arbor, Michigan*, 1717–28.

Hoffer, R. M., S. C. Noyer and R. P. Mroczynski 1978. A comparison of Landsat and forest survey estimates of forest cover. *Proc. Am. Soc. Photogramm., Fall Meeting, Albuquerque, New Mexico*, 220–31.

Hoffer, R. M. and staff 1975. *Computer aided analysis of Skylab multispectral scanner data in mountainous terrain for land use, forestry, water resource and geologic applications*. LARS Inf. Note 121275.

Holben, B. N. and C. O. Justice 1979. *Evaluation and modelling of the topographic effect on the spectral response from nadir pointing sensors*. NASA/GSFC Tech. Mem. 80305.

Holben, B. N. and C. O. Justice 1980. *An examination of spectral band ratioing to reduce the topographic effect on remotely sensed data*. NASA/GESFC Tech. Mem. 80640.

Horn, B. K. P. and R. J. Woodham 1978. *Destriping Landsat images*. MIT Artificial Intelligence Laboratory, AI memo no. 467.

Horn, B. K. P. and R. J. Woodham 1979. Landsat MSS coordinate transformations. *Proc. 1979 Symp. on Machine Processing of Remotely Sensed Data, Purdue University, Indiana*, 59–69.

Justice, C. O. 1978a. *The effects of ground conditions on Landsat multispectral scanner data for an area of complex terrain in S. Italy*. Unpub. PhD Thesis, Reading University, UK.

Justice, C. O. 1978b. An examination of the relationships between selected ground properties and Landsat MSS data in an area of complex terrain in southern Italy. *Proc. Am. Soc. Photogramm., Fall Meeting, Albuquerque, New Mexico*, 303–28.

Justice, C. O., D. F. Williams, J. R. G. Townshend and R. A. G. Savigear 1976. The evaluation of space imagery for obtaining environmental information in Basilicata Province, S. Italy. In *Land use studies by remote sensing*, W. G. Collins and J. L. van Genderen (eds), 1–39. Remote Sensing Society.

Kirby, M. E. and D. Steiner 1977. A theoretical model for the evaluation of the interactions between Landsat MSS data and UTM maps in geometric transformations. In *Remote sensing of Earth resources* **6**, F. Shahrokhi (ed.), 407–21. Tullahoma, Tennessee.

Labrandero, J. L. 1976. Madrid and its environment: land use and soils by digital analysis of Landsat data. In *Thematic mapping, land use, geological structure and water resource in central Spain*, N. De las Cuevas (ed.), 128–53. NASA Final Rep. E77–10167.

Lenco, M. 1978. *Exploitation d'une scène Landsat sur le Languedoc zones humides et couvert végétal*. Operation Pilote Interministerielle de Télédetection, Paris.

Lenco, M., Y. Heymann and P. Ferrault 1978. *Experience de télédétection aérienne sur la vallée de la Garonne et la région de Montauban.* Ministère de l'Environment et du Cadre de Vie, La Documentation Française, Paris.

Le Toan, T., P. Cassirame, J. Quach and R. Marie 1977. Inventory of ricefields in France using Landsat and aircraft data. *Proc. 11th Int. Symp. on Remote Sensing of Environment, Ann Arbor, Michigan,* 1483–95.

Long, G. 1977. The Écotheque Méditerranéenne. *Nature and Resources* XIII, no. 4.

Miller, L. D., K. Nualchawee and C. Tom 1978. *Analysis of the dynamics of shifting cultivation in the tropical forests of northern Thailand using landscape modelling and classification of Landsat imagery.* NASA Tech. Mem. 79545.

Miller, L. D. and D. L. Williams 1978. Monitoring forest canopy alteration around the world with digital analysis of Landsat imagery. *Proc. ISP/IUFRO Int. Symp. on Remote Sensing Observation and Inventory of Earth Resources and the Endangered Environment, Freiburg* 1721–61.

Morandini, R. 1977. Problem of conservation management and regeneration of Mediterranean forests: research priorities. In Unesco (1977), 73–9.

Mountjoy, A. B. 1974. The Mezzogiorne. In *Problem regions in Europe,* D. I. Scargill (ed.). Oxford: Oxford University Press.

NAS (National Academy of Sciences) 1977. *Resource sensing from space: prospects for developing countries.* Washington: NAS.

NASA 1975. *Skylab earth resources data catalog.* NASA–TM–X–70411.

NASA 1976. *Landsat data users handbook,* revised edn. Greenbelt, Md: Goddard Space Flight Center.

Naveh, Z. 1971. The conservation of ecological diversity of Mediterranean ecosystems through ecological management. In *Management of animal and plant communities for conservation,* E. Duffey and A. S. Watts (eds), 600–25. Oxford: Blackwell.

Nunnally, N. 1974. Interpreting land use from remote sensor imagery. In *Remote sensing: techniques for environmental analysis,* J. E. Estes and L. W. Senger (eds), 189–225. California: Hamilton Publishing Company.

OECD (Organisation of Economic Cooperation and Development) 1969. *Agricultural development in southern Europe.* Paris: OECD.

OECD 1972. *Annual survey: Italy 1971–1972.* Paris: OECD.

OECD 1976. *Regional problems and policies in OECD countries,* vol. 1. Paris: OECD.

ORI 1978. *Multispectral linear array sensor development: summary report.* Prepared for NASA/Goddard Space Flight Center, June 1978.

Paskoff, R. P. 1973. Geomorphological processes and characteristic landforms in the Mediterranean regions of the world. In *Mediterranean type ecosystems: origins and structure,* F. di Castri and H. A. Mooney (eds), 53–9. New York: Springer-Verlag.

Peplies, R. W. and H. F. Keuper 1975. Regional analysis. In *Manual of remote sensing,* R. G. Reeves (ed.), 1947–98. Virginia: American Society of Photogrammetry.

Quezell, P. 1977. Forests of the Mediterranean basin. In Unesco (1977), 9–33.

RCA 1977. *Return Beam Vidicon (RBV) panchromatic two-camera system for Landsat-C.* Final rep. AE R–4231. Princeton, NJ: RCA Astro-electronics.

Reeves, R. G. 1975. *Manual of remote sensing.* Virginia: American Society of Photogrammetry.

Rey, P. A. 1974. Expérience Arnica/ERTS–1 résultats et perspectives. *Eur. Earth Resources Satellite Experiments, Frascati, Italy,* B. T. Battrick and N. T. Duc (eds), ESRO SP–100, 325–48.

Rifman, S. S. 1973. Digital rectification of ERTS multispectral imagery. *Symp. on Significant Results from ERTS–1, Goddard Space Flight Center,* 1131–42.

Rohde, W. G. 1978. Improving land cover classification by image stratification. *Proc. 12th Int. Symp. on Remote Sensing of Environment, Ann Arbor, Michigan,* 729–42.

Sabins, F. F. 1978. *Remote sensing: principles and interpretation.* California: Freeman.

Sadowski, F. G. and W. A. Malila 1977. *Investigation of techniques for inventorying forested regions. 1: Reflectance modelling and empirical multispectral analysis of forest canopy components.* Environmental Research Institute, Ann Arbor, Michigan.

Salomonson, V. V. 1978. Landsat–D, a systems overview. *Proc. 12th Int. Symp. on Remote Sensing of Environment, Ann Arbor, Michigan*, 371–85.

Scargill, D. I. 1974. *Problem regions in Europe.* Oxford: Oxford University Press.

Schnetzler, C. C. and L. L. Thompson 1979. Multispectral Resource Sampler: an experimental satellite sensor for the mid-1980s. *Proc. Soc. Photo-instrumentation Engineers Symp.* **183**, *Huntsville, Alabama*, 255–62.

Schrumpf, B. J. 1975. *Multiseasonal-multispectral remote sensing of phenological change for natural vegetation inventory.* Unpubl. PhD Thesis, Oregon State University.

Strahler, A. H., T. L. Logan and N. A. Bryant 1978. Improving forest classification accuracy from Landsat by incorporating topographic information. *Proc. 11th Int. Symp. on Remote Sensing of Environment*, Ann Arbor, Michigan, **2**, 927–42.

Tom, C., L. D. Miller and J. W. Christenson 1978. *Spatial land use inventory, modelling and projection, Denver Metropolitan area with inputs from existing maps.* NASA/GSFC Tech. Mem. 79710.

Tomaselli, R. 1977. Degradation of Mediterranean maquis. In Unesco (1977), 33–73.

Townshend, J. R. G. and C. O. Justice 1980. Unsupervised classification of MSS Landsat data for mapping vegetation in an area of complex terrain: principles and problems. *Int. J. Remote Sensing* **1**, 105–20.

Townshend, J. R. G., D. F. Williams and C. O. Justice 1979. An evaluation of Landsat 3 RBV imagery for an area of complex terrain in S. Italy. *Proc. 13th Int. Symp. on Remote Sensing of Environment, Ann Arbor, Michigan*, 1839–52.

Tucker, C. J. 1978. A comparison between the first four thematic mapper reflective bands and other satellite sensor systems for vegetational monitoring. *Proc. Amer. Soc. Photogram., Fall Meeting, Albuquerque, New Mexico*, 579–93.

Unesco 1977. *Mediterranean forests and maquis: ecology, conservation and management.* Man and the biosphere, Tech. Notes 2.

Unesco–FAO 1963. *Bioclimatic map of the Mediterranean region 1:5 000 000 and explanatory notes.* Paris: Unesco.

van Genderen, J. L. and B. F. Lock 1977a. Testing land use map accuracy. *Photogramm. Engng and Remote Sensing* **43**, 1135–7.

van Genderen, J. L. and B. F. Lock 1977b. *A methodology for producing small scale rural land use maps in semi arid developing countries using orbital imagery,* NASA Final Rep. CR 151173, investigation no. SR9686, Department of Industry, London.

Vincent, R. K. 1973. Ratio maps of iron ore deposits Atlantic City district, Wyoming. *Proc. Symp. of Significant Results obtained from ERTS–1, Goddard Space Flight Center* **1**, 379–86.

Vita-Finzi, C. 1969. *The Mediterranean valleys: geological changes in historical times.* Cambridge: Cambridge University Press.

Williams, R. S., Jr and W. D. Carter 1976. *ERTS–1: a new window on our planet.* US Geol. surv. paper 929.

Williams, D. L. and G. F. Haver 1976. *Forest land management by satellite: Landsat-derived information as input to a forest inventory system.* Greenbelt, Md: NASA/Goddard Space Flight Center.

Winch, K. L. 1976. *International maps and atlases,* 2nd edn. London: Bowker.

Yassoglou, N. J., E. Skordalakis and A. Koutalos 1973. Application of ERTS–1 imagery to land use, forest density and soil investigation in Greece. *Proc. 3rd ERTS–1 Symp., NASA/Goddard Space Flight Center* **1**, 159–82.

153

7 Integrated land survey methods for the prediction of gully erosion

David F. Williams

7.1 Introduction

In many areas of the world, water-induced soil erosion is one of the major hazards resulting from changes in land use. Indiscriminate land development, particularly in semi-arid areas can have serious long-term effects. One component of the problem is that once gullying develops, it dissects the land and renders it unsuitable for further agricultural use unless expensive remedial measures are undertaken. Thus the cost of development is lost, whilst to maintain production, new areas must be developed at further expense. A second component is that the eroded material may be redeposited downslope, either on fertile flood plains or more seriously in reservoirs which are built to provide water for domestic and agricultural needs (Rapp *et al.* 1972). Reservoir infill can be a major problem as the entire region which is dependent upon the reservoir for its water supply is affected. Moreover, gully erosion is often indicative of accompanying sheet, splash and rill erosion, which in terms of the total sediment eroded may exceed that of gullying.

In view of the damage that gully erosion can cause, it is important that the potential extent and severity of future gully erosion is delimited in the early planning stages of land development schemes. Generally, three distinctive scales or intensity of land survey can be recognized. The reconnaissance survey is usually carried out at country or regional levels at scales of 1:2000000 or smaller. At this scale the major aim is to provide an inventory defining areas where a more detailed survey should be made. The semi-detailed survey is usually undertaken at scales of 1:25000–1:50000, the aim of such surveys being to assess the suitability of land for different land uses, the need for irrigation development, dams and roads. At the most detailed scale (1:10000 and larger), there are development surveys which involve assessment of land capability at the farm or field level as well as variables such as estimated future crop yields, required fertilizer levels, equipment needs, and the overall economic viability of recommended changes.

Assessment of the consequences of erosion should not be left to the development survey stage when, for example, the harmful effects of individual gullies need to be evaluated, but instead should be investigated both at the semi-detailed and reconnaissance scales. At the latter scale areas where gullying is currently a serious problem should be recognized and delimited for further investigation. At the semi-detailed level it should be possible to delimit areas potentially susceptible to gully erosion and to describe those land use practices exacerbating its development. If possible rates of extension should also be estimated to enable determination of sediment loss to be made.

154

Huntingdon (1914) proposed the 'trigger' concept of gully initiation, which is based on the idea that relatively small changes in environmental conditions can result in very significant increases in erosion rates. In more modern terminology we might say that a system may pass across a threshold due to a small change in one critical variable and force the system radically to adjust itself in order to achieve a completely different steady state (Strahler 1958).

Changes in climate have frequently been held responsible for initiating gullying (e.g. Huntingdon 1914, Bryan 1925, 1941, Vita-Finzi 1969, Cooke & Reeves 1976). Scarcity of long-term records has often made this difficult to disentangle from effects of expected but rare rainfall events. Climatic change is also inextricably bound with concomitant vegetational change.

Over shorter time spans, in terms of land capability assessment, those vegetation changes produced by man are of principal importance. Changes in land cover type undoubtedly often lead to gullying as is indicated by the association which is frequently found between arable land and gully erosion (e.g. Lyell 1872, Ireland *et al*. 1939, Nir & Klein 1974). Changes in the intensity of land use activities can also be responsible for gullying. The most commonly quoted example is overgrazing (e.g. Rich 1911, Cotton & Stewart 1940, Hastings 1959, Cooke & Reeves 1976) where the decrease in vegetation cover leads to increases in run-off, possibly accompanied by decrease in soil strength and thus in increased erosion. Blong (1966) additionally pointed to the importance of stock trampling in compacting the surface, which hence increases run-off.

Although changes in climate and land cover are undoubtedly critical in gully initiation their significance will be heavily modulated by the local conditions of relatively unchanging properties such as those of topography, pedology and geology.

7.3 Framework for the survey

A comprehensive forecast of the future extent of gullying is comprised of two separate tasks. One involves prediction of the future extent of existing gullies by headward extension and bifurcation creating new branches. Such prediction usually demands a consideration of the local terrain conditions immediately adjacent to the gullies and hence requires a detailed (large scale) survey. A methodology for tackling this problem has been proposed based primarily on considerations of slope form relative to characteristic gully long profiles in different lithologies (Williams 1978). In contrast, this chapter concerns itself with a different task namely locating those areas as yet ungullied, but which are vulnerable if environmental conditions change. The geographical scale of such a survey is inevitably much smaller than the previous one since the whole of a region needs to be considered and not merely those areas adjacent to existing gullies. Furthermore separation of the two types of survey is justified in geomorphological terms, since those terrain characteristics important in initiating gullying are often different from those leading to its further extension: for example, once initiated, a gully system may readily extend into areas with a vegetation cover much too dense to have permitted initiation there.

Three main alternative approaches can be recognized for locating areas vulnerable to gullying.

Direct experiment. Within the area of study, plots of land could be selected, land cover changes made and the development of future gullying monitored. However, it is probable that a large number of sites would have to be selected and conceivably it could be several years before any erosion occurred to enable the land to be classified. Thus the approach is unsuitable within the constraints of terrain evaluation schemes where results are normally required in a short period of time.

Sequential air photography. This can be used to recognize, monitor and explain the development of new gullies. The results can then be used for prediction to analogous units of land. Where photographs are available and if suitable land cover changes had occurred this approach could be successful. However, there are large areas of the world, and in particular the developing world, where suitable multidate photography is not available.

Figure 7.1 Flow diagram showing major stages in prediction of areas susceptible to gullying.

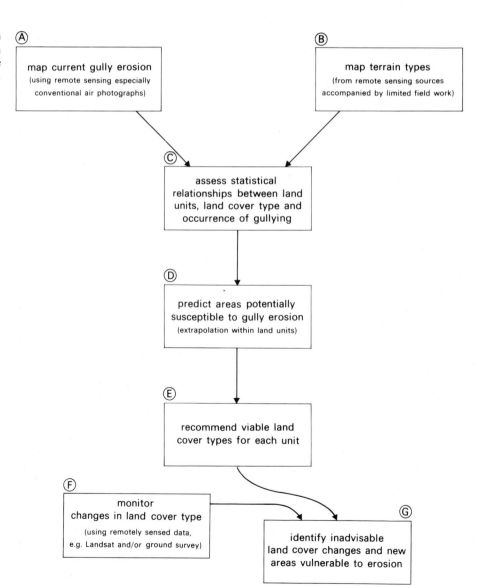

Terrain classification. In this approach the area is classified into its terrain units and these are related statistically to the occurrence of gullying, thus enabling the factors important in gully development to be identified. The results can then be used for prediction between analogous units of land. This latter approach is developed here, since it is least demanding of ground survey and remote sensing data. There are numerous examples of terrain classification and evaluation methods (e.g. Stewart 1968, Mitchell 1973, Young 1976), some of which are reviewed in Chapters 1 and 5 of this book. Young, in his review, points out the general inadequacy of existing methods in a predictive role. Their major limitation is that terrain evaluation is treated as a 'static' subject; that is, it is assumed that the terrain can be assessed in terms of the conditions prevailing at the time of survey. However, he points out that a 'dynamic' situation normally exists inasmuch as the intended future land usage needs to be considered with respect to its possible beneficial or adverse effect on the quality of the land.

Nevertheless, it should prove possible to adapt one of the existing methods of terrain classification and include consideration of changes in land cover types in the analysis. The major advantages of this type of approach would be its general applicability and speed of execution. The principal stages of our approach are outlined in Figure 7.1 and are described in more detail in the remainder of this chapter. At present the relationship between gullying, terrain conditions and land cover types is based primarily on their statistical association rather than on specific physical models of gully developments. Incorporation of the latter is clearly highly desirable in the future assuming that they rely on data which can be collected relatively inexpensively. Of course the present choice of the specific terrain characteristics in the present empirical study are dependent upon existing knowledge of gully behaviour.

7.4 The study area

The study area is located within Basilicata Province, southern Italy and is drained by the rivers Agri and Sinni and their tributaries (Fig. 7.2). Gully erosion is a severe problem within the area, and its distribution is principally due to lithology, landforms, land cover types and land management. Geologically much of the area is very young, consisting mainly of Plio-Pleistocene sands, gravels and marine clays. The pre-Pliocene deposits are mainly limestones, micaceous mudstones, sandstones and conglomerates which have been heavily folded and faulted. Post-Pliocene uplift of at least 800 m has resulted in areas of high relief which are inherently unstable. Due to extensive clearance there is now very little natural vegetation, with relatively small areas of woodland, and large areas of grazed shrubland and cultivated land.

Within the area shown in Figure 7.2 three sample areas each of 64 square kms were selected to test the approach. They contain a wide range of lithologies and terrain types within which gullying is known to occur at the present time. For each of the three areas the extent of the current gully erosion was mapped from 1:17000 air photographs taken in 1974. To assess the accuracy with which the gullies were identified, 108 gullies were randomly selected using the air photographs and a field check undertaken. On checking, all but six of the gullies were found to exist giving an accuracy of 94%. This 6% represents an error of commission. No assessment of the error of omission can be made without a complete field survey but extensive field observation suggests it is very small and involves only minor gullies. The surface

157

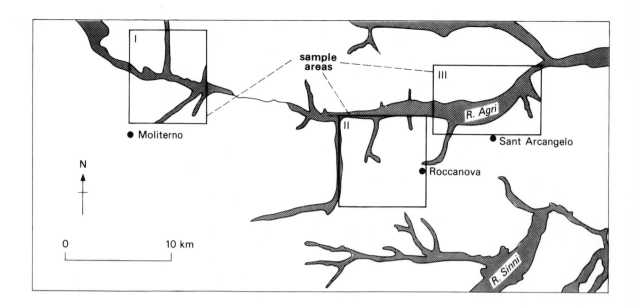

Figure 7.2 Location of
sample areas. areal extent of gullying was mapped instead of individual gullies, because in areas of
severe erosion individual gully segments could not be recognized on the photographs and in small gullies it was impossible consistently to recognize the cross
sectional form and hence the size of gullies.

7.5 The prediction of areas susceptible to gully erosion

Two principal methods of areal classification are possible, namely the physiographic and parametric, both of which are described in outline in Chapter 5. Objective comparisons between the two approaches have been carried out (Williams
1978). It was found that unless a very small cell size was used the former was more
successful in defining homogeneous terrain units. Use of a sufficiently small cell size
would have led to unreasonably high data collection demands in the parametric
approach: in particular, a high redundancy of data would have resulted in areas with
a low frequency of gully erosion and in areas with uniform terrain.

Table 7.1 Definition and terminology of physiographic terms.

Name	Definitions
Land system	A recurrent pattern of genetically linked land units
Land unit	One or more land elements grouped: an area of land which for practical purposes is reasonably homogeneous or for which a change in any one property may be recognized. Diagnostic features are: slope morphology, lithology, relief amplitude, origin, processes.
Land component	An area of land within a land unit which has one dominant land cover type. One or more land elements grouped in terms of their land cover type.
Land element	Simplest part of the landscape; for practical purposes has uniform lithology, form, soil and vegetation.

In this study a methodology which is essentially physiographic has been used (Ch. 5). Such an approach inherently implies that the land surface can be readily divided into a finite number of units having natural boundaries. Furthermore, the assumption is made that analogous units may be recognized, which enables prediction between these units. The methodology is similar to that defined by Brink *et al.* (1965) and the definitions of the terms are shown in Tables 7.1 and 7.2. The main variations from Brink are the introduction of the term land component, which is an area of land within a unit where there is only one land cover type. For the purpose of this study, which requires that the land cover type be recognizable on black and white air photographs at scales of 1:15 000 to 1:30 000, only six categories of land cover types were used (Table 7.2).

For each of the three sample areas in Figure 7.2 the land systems and units were mapped. If any part of a land unit was found to be gullied, then throughout the whole area it was subdivided at the land component level. This was to enable the land cover types associated with gully erosion to be identified. The resultant map of part of area 1 is shown in Figure 7.3. The statistical relationships between the occurrence of gully erosion and the land systems, units and components have been established by the construction of contingency tables (Tables 7.3, 7.4 and 7.5). In Table 7.4, land units are only included if they belong to land systems possessing at least some gullying. Similarly in Table 7.5, land components are only included if gullying was present in at least one occurrence of the land unit. All those land systems suffering from gullying are underlain by unconsolidated or poorly consolidated materials. However, no precise information on where gullying occurs within a system can be obtained because of the wide variety of landforms present within each system, and thus it was decided to work at the land unit level.

Table 7.2 Variables used in definition of land systems, units and components.

Land systems
Type and occurrence of land units (see below) with particular emphasis placed on structural boundaries and lithology.

Land units
(1) geology – see above
(2) morphology – slope angle, slope length, relief amplitude, slope form (convex, concave, linear); aspect; plan concavity; plan convexity.
(3) geomorphology – landform types, (especially structural landforms and depositional landforms).
(4) surface materials – presence/absence, depth, stoniness, texture.

Land components
Land unit properties plus land cover types (only dominant cover was considered):
(a) woodland
(b) shrubs
(c) herbaceous – permanent (pasture)
(d) cultivated – seasonally bare
(e) barren – perennially bare

159

Key

land system 1

land system 2

land system 3

areas of gullying

———— land system boundary

——— land unit boundary

----- land unit – component

A land unit

c land unit – component

0 _____ 2 km

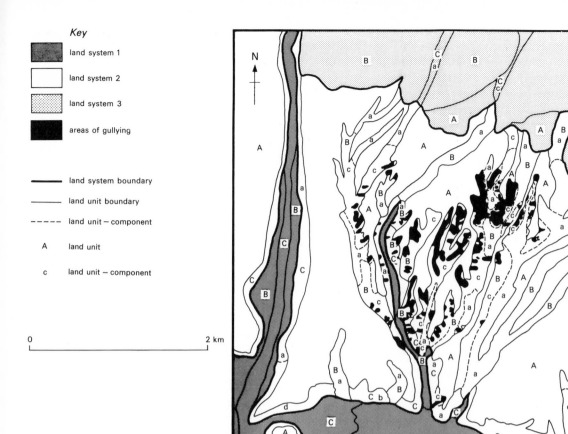

Figure 7.3 Land unit map with existing gully network.

Table 7.3 Contingency table relating the occurrence of gullying to occurrence of land systems.

Land system	Lithology	Gullies present	Gullies absent	% of total area mapped
1	Sandstone	1	1	2·2
2	Conglomerate over clay	5	3	23·2
3	Alluvial deposits	0	7	20·0
4	Limestone	0	1	2·4
5	Sandstone	0	5	0·6
6	Sands	2	1	6·5
7	Conglomerates	3	0	13·6
8	Consolidated conglomerates	0	1	0·7
9	Conglomerates	1	0	10·1
10	Sands	2	0	2·1
11	Flysch	0	2	3·4
12	Clay	2	3	15·2

160

Table 7.4 Contingency table showing the occurrence of gullies within units, for land systems within which gullying occurs.

Land unit	Number of units		% of total area
	gullies present	gullies absent	
1A	0	4	0·7
1B	0	4	1·5
1C	1	1	0·1
2A	0	17	16·0
2B	15	11	3·8
2C	8	10	3·4
6A	3	0	2·4
6B	2	8	1·6
6C	1	3	0·8
6D	0	5	0·1
6E	0	1	1·8
6F	0	3	0·4
7A	0	3	2·4
7B	4	8	2·2
7C	12	5	1·8
7D	5	10	5·9
7E	2	2	0·4
7F	0	4	0·7
9A	0	5	3·3
9B	4	5	1·8
9C	0	5	0·7
9D	13	9	3·7
9E	0	1	0·6
10A	1	11	1·1
10B	0	8	0·4
10C	0	3	0·3
10D	1	0	0·4
12A	20	11	3·8
12B	10	30	4·9
12C	3	0	2·6
12D	0	4	1·2
12E	0	6	1·0
12F	0	4	0·4
12G	0	3	1·2
Totals	108	215	

% of area defined by units within which gullying occurs is 39·2%. Those systems with no occurrences of gullying are excluded.

Gullying occurs within seventeen different types of land units out of a total of thirty-four, but on average only 33% of the occurrences of these land units actually contain gullying. In contrast, gullying is found in twenty-five out of a theoretical total of 170 types of land components, and 82% of all occurrences of the twenty-five types, actually contain gullies. Furthermore, 94% of all gullies occur in twenty different land components. In the mapped areas, gullying is mainly confined to three of the six land cover types, namely arable, pasture and barren (i.e. permanently bare ground). There are a few cases of gullying found within woodland but in every case this was due to extension of the gullies from adjacent non-woodland areas. Thus in summary it appears that the occurrence of gullying is poorly associated with particular land systems, but is moderately well associated with distinctive land units and is closely associated with particular land components (Table 7.5).

Table 7.5 Contingency table showing the occurrence of gullies within land components.

Land components	Number of components		% with gullies	Total
	gullies present	gullies absent		
1Ca	0	1	0	1
1Ca	1	0	100	1
2Ba	3	22	12	25
2Bc	24	3	89	27
2Bd	0	5	0	5
2Be	4	0	100	4
2Ca	3	21	12	24
2Cc	8	2	80	10
2Cd	0	9	0	9
6Aa	0	10	0	10
6Ac	7	0	100	7
6Ad	0	1	0	1
6Ba	0	6	0	6
6Bb	0	2	0	2
6Bc	1	1	50	2
6Bd	0	8	0	8
6Ca	0	2	0	2
6Cb	0	1	0	1
6Cd	1	5	17	6
7Ba	1	15	6	16
7Bc	7	2	78	9
7Ca	3	12	20	15
7Cc	13	3	81	16
7Da	1	17	6	18
7Dc	8	2	80	10
7Ea	0	2	0	2

Table 7.5 – *continued.*

Areas susceptible to gulley erosion

Land components	Number of components		% with gullies	Total
	gullies present	gullies absent		
7Ec	1	1	50	2
7Ed	2	0	100	2
9Ba	0	13	0	13
9Bc	6	1	86	7
9Bd	0	1	0	1
9Da	0	29	0	29
9Dc	15	2	88	17
9Dd	0	2	0	2
10Aa	0	3	0	3
10Ab	0	4	0	4
10Ac	1	2	33	3
10Ad	0	5	0	5
10Db	0	1	0	1
10Dd	1	0	100	1
12Ae	20	11	65	31
12Ba	0	3	0	3
12Bb	0	4	0	4
12Bc	20	3	87	23
12Bd	0	60	0	60
12Cc	6	0	100	6
12Cd	17	0	100	17
Totals	174	297		471

Only those land units which have at least one occurrence of gullying are included.
See Table 7.2 for meaning of letters a–e in labelling of land components.

Nevertheless, of the land component types which contain gullying, an average of 18% of their occurrences do not contain gullying at present. Explanation of this awaits a more fundamental understanding of the processes creating gullying which is beyond this current statistical approach. Three possibilities exist: firstly that the most appropriate land characteristics were not used or the best method of areal subdivision has not been applied; secondly, that the relevant environmental factors are unrecognizable when data collection is from air photographs with only small amounts of ground survey; thirdly, that these areas are liable to be gullied in the future with their existing land cover. Explanation for their present ungullied state may simply result from insufficiently high rainfall intensities having occurred locally. Under their current land cover, probably all three explanations are correct, but at the semi-detailed scale of survey it is advisable to assume these areas *are* vulnerable and should be subject to more detailed investigations if the consequences of their erosion are thought serious.

Working only at the land unit level we could conservatively predict that if gullying occurred in any one occurrence of a unit then all other occurrences are deemed vulnerable. On this assumption 39·2% of the total area mapped may be classified as potentially susceptible to gully erosion under existing or changed land cover.

The major limitation of this approach is that within a land unit the prediction of susceptibility to erosion can only be made if all possible land cover types and hence land components are present and this is not always found. To overcome this problem the concept of *abstract land units* and hence *analogous land units* between systems can be introduced. An abstract land unit is an idealized unit to which land units may be related and was first introduced by Perrin and Mitchell (1969). The major criterion for grouping the units together under one abstract unit is surface morphology. This is because land cover has not been included as definitive of units whilst lithology and structure were of major importance in defining systems. Obviously the grouping of units leads to a degree of generalization; for example, the slope classes are broader than for each individual land unit. In total six abstract land units have been recognized within which gullying occurs in at least one occurrence (Table 7.6). In terms of wider application than the present study, it should be noted that analogous units can only be recognized between areas relatively close together which have similar climate conditions (Mitchell 1969). Mitchell did not define how 'close' the areas should be, but here it has been possible to relate land units 50 km apart.

Table 7.6 Land cover types identified as being important in determining gully erosion within each abstract land unit.

Abstract land unit	Land use/cover type which will result in gullying	Land cover type which will not result in gullying
I	c,e	a,d(b)
II	c,e	a,d(b)
III	c (e)	a,b,d
IV	c,d(e)	a,b
V	c,d(e)	a,b
VI	c (e)	a,b,d

See Table 7.2 and text for key.

Table 7.6 indicates for each land unit and abstract land unit those land cover types for which it is predicted gully erosion will develop. The areal extent of these are indicated in Figure 7.4. Despite creation of the analogous units all cover types are still not found for each unit. However, at least in the case of category e (permanently bare ground) known to be the most readily gullied type, it is safe to assume that if present it would certainly lead to gullying for abstract units III, IV, V and VI. In the case of shrubland on units I and II, since this land cover is free of gullying in types III to VI, it is safe to assume that it is unlikely to lead to gullying.

It is interesting to note arable land has only been found to be associated with gullying within certain land units. This was an unexpected conclusion which is partly ascribed to the cover provided by crops for some of the most erosional times of year. But of greater importance is the frequent ploughing out of rills every year, which prevents the development of the more serious fluvial forms of erosion, on lower angle slopes. At least in the clay lands, the steeper cultivated slopes nevertheless suffer from erosion, but it is manifested in the form of shallow landslides.

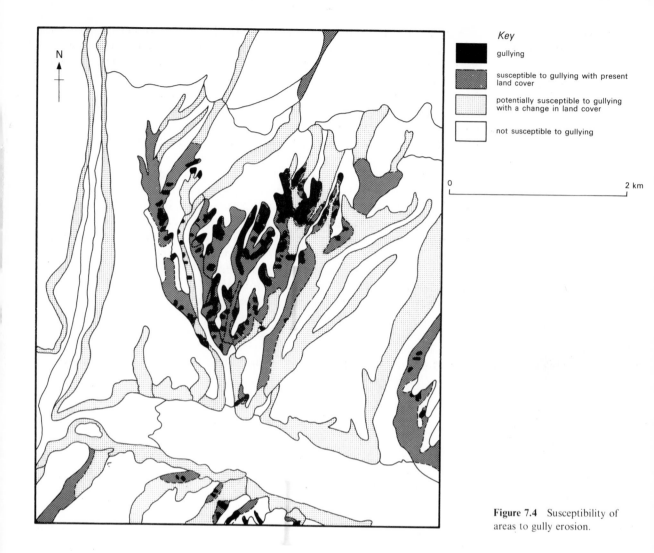

gullying

susceptible to gullying with present land cover

potentially susceptible to gullying with a change in land cover

not susceptible to gullying

0   2 km

Figure 7.4 Susceptibility of areas to gully erosion.

7.6 Conclusions

At the reconnaissance level, the physiographic approach may be used broadly to define areas where gully erosion exists and where it will be a problem in the future. But the precision with which such areas can be located will not usually be high since units with very different susceptibilities to gullying may be found within one land system.

At the scale of the semi-detailed land resource survey which is normally carried out at 1:50000 or less, it is possible to use the physiographic approach to predict areas susceptible to gully erosion at the land unit level and to identify broad land cover changes which may result in gully development. The methodology devised and applied in this chapter incorporates the 'dynamic' concept as envisaged by Young, in that the future intended land use is taken fully into account. The method distinguishes clearly between the relatively static terrain features such as lithology, slope and the more rapidly changing feature, namely the land cover type. Additionally, it should be recognized that the relationships between terrain conditions, land cover type and gullying may be affected by climatic change.

165

Within the framework of a natural resource survey, information on units susceptible to gully erosion will enable these units to be designated for conservation practice. In addition, it should be realized that it is not just the immediate unit but the entire drainage catchments related to those units which need to be conserved.

A major advantage of using a physiographic approach is that the information can be acquired for relatively large areas, in a relatively short period of time and at an early stage within a resource survey. This is important because information is often required quickly, which prohibits long-term studies prior to planning and obviously the more information that can be obtained at an early stage then the more accurate and better will be the eventual plans. However, solving the problem of gully erosion is but one aspect of a development scheme. The importance attached to any gully erosion problem must be viewed within the framework of a resource survey as a whole, and involves consideration of economic and social aspects as well as the land capability. Thus in areas where gullying is only a minor problem in terms of the area affected, it may not be worthwhile to make a detailed analysis of the problem.

In the area studied in this chapter the results of the analysis presented have an immediate applicability. The area has no regional planning scheme designed to rationalize the agricultural practice. This has resulted in no control over land use changes or grazing and has led to widespread gullying and destruction of land. In addition there are three dams within the area which are used to supply water for irrigation projects. One of these dams built only 23 years ago has already silted up and is now virtually useless as a source of irrigation water, being used solely as a flood control barrier. If land use recommendations based on the methods discussed in this chapter were applied within the catchments of these dams, then the problem of siltation could be greatly reduced, whilst at the same time the agricultural practice could be improved.

7.7 Looking to the future

At the time this study was initiated in 1974, the major source of data for integrated surveys was black and white air photographs. The principal constraint of such imagery is that each photograph only covers a very small area thus making it difficult to recognize analogous land systems and units over even relatively small areas. This problem has normally been overcome by the use of photo-mosaics. However, unless orthophotos are used then the resultant mosaic is cartographically very inaccurate. Satellite imagery from the Landsat series has been available for several years but its major limitation has been that of coarse spatial resolution; Landsat 3 $\overline{\text{R B V}}$ imagery now offers a partial answer to this problem. This imagery has a nominal ground resolution of 24 m which compares favourably with the resolution of 79 m of the MSS imagery. With such a resolution coupled with high cartographic accuracy the imagery overcomes the main criticisms of the MSS data, which was that features could not be consistently nor accurately located in areas of high terrain diversity.

The area shown in Figure 7.3 is reproduced in Figure 7.5 which is an R B V image. Although the resolution is still not quite sufficient for the production of a comparable terrain map, the principal changes of land cover can undoubtedly be monitored. The R B V imagery is monospectral and only where land cover types contrast strongly with their surroundings, as in the case with the oak woodland in this area, can land cover be mapped and changes monitored (Townshend *et al.* 1979). Satellite-based multispectral systems such as Landsat D's thematic mapper

Figure 7.5 Landsat-3, return beam vidicon image of area depicted in Figure 7.3. The dark areas with the exception of the reservoir represent areas of oak woodland.

and the French SPOT system (Table 2.3), with comparable resolution to the RBV imagery, will be operating by the mid-1980s. Their discriminating ability between land cover types should be substantial and this will greatly help the prediction of future gully erosion. Nevertheless, the enormous volume of data to be processed from those future systems will hinder their application without considerable improvements in image processing systems.

References

Blong, R. J. 1966. Discontinuous gullies on the volcanic plateau. *J. Hydrol. (NZ)* **5**, 87–99.

Brink, A. B. A., J. A. Mabbutt, R. Webster and P. H. T. Beckett 1965. *Report of the working group on land classification and data storage.* MEXE rep. no. 940.

Bryan, K. 1925. Date of channel trenching (arroyo-cutting) in the arid southwest. *Science* **62**, 344–88.

Bryan, K. 1941. Physiography. *Geol. Soc. Am. 50th Ann. Vol.* 3–15.

Cooke, R. U. and R. W. Reeves 1976. *Arroyos and environmental change in the American south-west*. Oxford: Oxford University Press.

Cotton, W. P. and G. Stewart 1940. Succession as a result of grazing and of meadow desiccation by erosion since settlement in 1862. *J. Forestry* **38**, 613–26.

Hastings, J. R. 1959. Vegetation change and arroyo cutting in south-eastern Arizona. *J. Arizona Acad. Sci.* **1**, 60–7.

Huntingdon, E. 1914. *The climatic factor as illustrated in arid America*. Carnegie Inst. Publ. 192, Washington.

Ireland, H. A., C. F. S. Sharpe and D. H. Eargle 1939. *Principles of gully erosion in the piedmont of south Carolina*. USDA Tech. Bull. 633.

Lyell, C. 1872. *Principles of geology*, 11th edn. New York: Appleton and Co.

Mitchell, C. W. 1969. *An appraisal of physiographic units for predicting site conditions important to agriculture in arid areas*. Unpubl. PhD Thesis, University of Cambridge.

Mitchell, C. W. 1973. *Terrain evaluation*. London: Longman.

Nir, D. and M. Klein 1974. Gully erosion induced in land use in a semi-arid terrain (Nahal Shigma, Israel). *Z. Geomorph. Suppl.* **21**, 191–201.

Perrin, R. M. S. and C. W. Mitchell 1969. *An appraisal of physiographic units for predicting site conditions in arid areas*. MEXE rep. no. 1111, 2.

Piest, R. F., J. M. Bradford and R. G. Spomer 1975. Mechanism of erosion and sediment movement from gullies. In *Present and prospective technology for predicting sediment yields and sources*, USDA Agric. Res. surv. ARS-S, 40, 162–76.

Rapp, A., L. Berry and P. Temple 1972. Soil erosion and sedimentation in Tanzania – the project. In *Soil erosion and sedimentation in Tanzania*, A. Rapp, L. Berry and P. Temple (eds). *Geog. Annlr.* **54A**.

Rich, J. L. 1911. Recent stream trenching in the semi-arid portion of south-western New Mexico, a result of removal of vegetation cover. *Am. J. Sci.* **32**, 237–45.

Stewart, G. A. (ed.) 1968. *Land evaluation*. Melbourne: Macmillan.

Strahler, A. N. 1958. Dimensional analysis applied to fluvially eroded landforms. *Bull. Geol. Soc. Am.* **69**, 279–300.

Townshend, J. R. G., D. F. Williams and C. O. Justice 1979. An evaluation of Landsat 3 RBV imagery for an area of complex terrain in S. Italy. *Proc. 13th Int. Symp. on Remote Sensing of Environment, Ann Arbor, Michigan*, 1839–52.

Vita-Finzi, C. 1969. *The Mediterranean valleys*. Cambridge: Cambridge University Press.

Williams, D. F. 1978. *The identification and location and intensity of gully erosion in Basilicata Province, southern Italy*. Unpub. PhD Thesis, Reading University, UK.

Young, A. 1976. *Tropical soils and soil survey*. Cambridge: Cambridge University Press.

Reconnaissance land resource surveys in arid and semi-arid lands

Colin W. Mitchell

8.1 Arid and semi-arid lands: extent and distribution

Those parts of the Earth's land surface in the tropical and temperate zones have been defined and mapped for Unesco on the basis of Thornthwaite's aridity index by Meigs (1953). They are shown in Figure 8.1 and Table 8.1. They fall into three classes: semi-arid, arid and hyper-arid. Broadly speaking, the semi-arid areas have an annual rainfall which is less than $\frac{2}{3}$ of the potential evapotranspiration, arid areas have less than $\frac{1}{3}$ and hyper-arid areas have additionally recorded at least one 12-month period without rainfall. Practically, the semi-arid lands can be categorized where they are covered by short grassland without trees or *savanna* where the grass is longer and there are scattered trees. They are mainly used for dry farming and extensive ranching or nomadic stock rearing. The arid lands can be labelled *sahel* and have sparse scrubby vegetation suited only to the extensive grazing of cattle, sheep, and goats where water is obtainable from wells, springs, or rivers, and the hyper-arid Saharan areas will scarcely sustain life that is not based on the camel. Irrigation is the only interruption to this pattern and is essential to any kind of intensive agriculture in these zones.

Figure 8.1 Principal arid regions of the world (after Joly 1957).

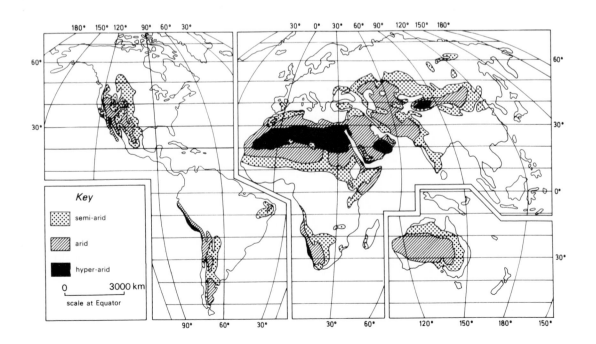

The importance of arid lands to food production in the modern world scarcely needs emphasizing. Table 8.1 shows that they occupy almost $\frac{1}{3}$ of the world's total land surface. They are important in all continents except Europe, dominant in Africa and Australia and especially widespread in the Third World. Their density of population is much sparser than that of the world as a whole, as Table 8.2 makes clear: within the arid and hyper-arid areas, almost $\frac{3}{4}$ of the population lives in the irrigated Nile, Tigris, Euphrates and Indus valleys.

Table 8.1 Proportion of land surfaces of continents in arid and semi-arid zones (after Joly 1957).

Continents \ Areas	Total area (km² × 10³)	% semi-arid	% arid	% hyper-arid	% total arid
N. and central America	22 162	10·6	6·7	1·9	19·2
S. America	17 755	9·0	7·5	2·0	19·5
Africa	29 797	18·5	24·5	15·0	58·0
Asia	42 365	1·5	19·0	3·0	37·0
Australia	7 704	29·0	51·0	0	80·0
Europe	10 032	7·5	2·0	0	9·5
Other (polar areas, Greenland, Indonesia, New Zealand, Oceania, etc.)	23 418	0	0	0	0
World	153 233	12·2	14·6	4·2	31·0

Table 8.2 Distribution of population in arid areas (from Joly 1957, Hills 1966).

	Population (millions)	% of world population	% of world land area	Average density of population: persons/km²
Semi-arid	276	9·2	12·2	13·0
Arid	103	3·4	14·6	4·7
Hyper-arid	5	0·2	4·2	0·8
Total	384	12·8	31·0	7·9 (world 19·6)

8.2 Information needs

Land resources in the broadest sense include soils, water, vegetation, minerals, natural sources of energy and the environment generally, the first three being especially critical for human life. Information about them is needed at a number of different scales: reconnaissance for broad planning, semi-detailed for identification of local development potentials and detailed for feasibility studies and project planning. The need for reconnaissance information tends to increase rather than decrease with the availability of detailed local studies because of the continuing requirement for a readily comprehensive correlative framework and for overviews of specific resource factors. Today it is urgently needed for forecasting and planning agricultural production for international trade and aid, and for foreseeing, warning, and preventing or mitigating natural calamities.

Although information about natural resources has increased rapidly of recent years and the media for its communication have improved, arid regions still remain relatively little known. There are a number of reasons for this. First, much of their extent is in the Third World while most of the information recently acquired relates to other areas already relatively well known, so that the gap in intelligence between the developed and the developing parts of the world is widening. Second, there are serious economic, social, and political restrictions on the acquisition of both national and international resource information. Questions of national defence, commercial competition and the multiplicity of intelligence-gathering organizations make it hard to obtain rapid overall assessments.

Nevertheless, there has recently been a vast increase in the amount of available data mainly in the form of literature, maps, and aerial imagery, much from orbital altitudes. This has led to another problem: that of data management. We are no longer able adequately to store, collate, retrieve or communicate them, far less act on them. In an increasing number of areas data management has replaced data acquisition as the main bottleneck in resources evaluation and this problem is becoming more general.

8.3 Data collection methods

Since land resource factors are fixed in location and extent they are amenable to geographical forms of analysis. There are now many types of thematic mapping in use which have been developed from the original geological, geomorphological, pedological, climatic, hydrological and vegetation bases. Geological maps based on stratigraphy for instance, can now be supplemented for example by materials, minerals, metallogenic, tectonic, Quaternary, hydrogeological, geohydrochemical and engineering geological maps; soil maps based on pedology can be supplemented by soil engineering, land capability, or soil degradation maps. Very few of these types are available for arid or semi-arid areas so that a simple thematic mapping base is highly desirable to form a correlative framework for the main resource factors.

Such a framework can be provided by an analysis and classification of the landscape as a basis for the 'pigeonholing' of local information. Landscape assessments, if based on interpretations of genesis and processes as well as observable surface attributes, can have predictive power through suggesting causation. This is the essential element of the 'systems' approach to geomorphology and makes it of particular value for integrated inventories of economically important land characteristics, notably soils, water, vegetation and land use. Landscape units, which control and explain their distributions and local interrelationships, can be defined and mapped. This gives an integrated approach to resource factors which provides a valuable basis for the assessment of land development priorities and likely hazards from erosion and other forms of degradation.

The basic units of landscape first defined by Christian and Stewart (1964) are *land systems* which are recognizably current patterns of landscape but which are not themselves internally homogeneous. The smaller units of which they are composed are known as *land facets* and it is these which are the practically important basic subdivisions of the landscape for most purposes. Recent research (Mitchell *et al.*

171

1979) has shown that more than half of the total variance of most land and soil properties within a whole climatic zone is associated with a subdivision at land system and land facet scales and almost $\frac{1}{3}$ at land facet scale alone so that wherever possible even reconnaissance landscape interpretations should be made at these scales.

The value of an integrating landscape basis is further underlined by the development of remote sensing from an increasing variety of sensors and platforms which provide information of many types and on many scales. For the reconnaissance of arid and semi-arid lands photography and other imagery from orbital altitudes are especially appropriate because of the clarity with which the surface can be interpreted in the relative absence of both cloud cover and dense vegetation. Such imagery includes film from rockets and manned space flights such as Gemini, Apollo, and Skylab and various forms of imagery from unmanned satellites such as Landsat, Seasat, Meteosat, and possibly in the future, Stereosat (Table 2.3). Landsats 1, 2, and 3 have already provided a wide repetitive coverage of most of the Earth's surface with acceptably accurate resolution and coordinate location of detail. They provide the great majority of such imagery currently in use in small-scale land resources survey and the only platforms considered below.

8.4 Data analysis

The basic problems in the practical interpretation of all remotely sensed imagery have been considered in earlier chapters. Their essence is to recognize links between the phenomena being studied on the ground and the tones, grain and patterns of light (heat, or sound) which are recorded by the camera, scanner or other sensor carried in the aircraft or satellite. These can be considered as consisting of myriads of pixels which can either be recorded digitally on magnetic tape or printed in analog form as pictures.

The interpretation of Landsat imagery differs from that of conventional aerial photography in that the much smaller scale renders invisible many of the features familiar to the ordinary observer, such as small landforms, fields, houses, roads, trees, etc.. Thus it necessitates the evolution of a two-step method of interpretation. First, the broad scale visible attributes of the theme being mapped (geology, soils, vegetation, etc.) are identified, and all available information about them is derived from existing literature, maps, and field experience. Secondly, a scheme of Landsat interpretation is devised which comes as near as possible to delimiting these attributes. This involves simple visual interpretation of black and white or colour composite Landsat frames but is usually aided by the use of image enhancement procedures such as additive viewing or density slicing techniques on the analog imagery or by more sophisticated analytical methods using computer compatible tape, e.g. ratioing (Section 4.2). The basic interpretation problem becomes one of marrying the features observable on the imagery to the attributes required in mapping.

Three FAO projects have explored the application of Landsat to reconnaissance resource evaluations in arid and semi-arid areas and they illustrate the development of the methods and foreshadow the realization of their potential. These projects were: an experimental test of the usefulness of such imagery in the Sudan (Mitchell 1975), a land system evaluation of Jordan at 1:1 m (Mitchell 1978), and a first-stage input into the soil degradation map of the world at 1:5 000 000 (Mitchell & Howard 1978).

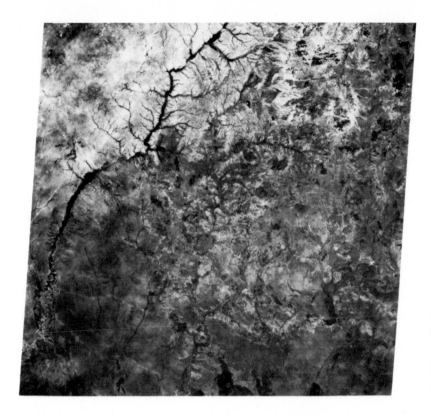

Figure 8.2 Landsat 1 frame covering Wadi El Ghalla, Nuba Mountains region of Kordofan, Sudan, MSS5 of 9 November 1972. Area depicted 185 × 185 km.

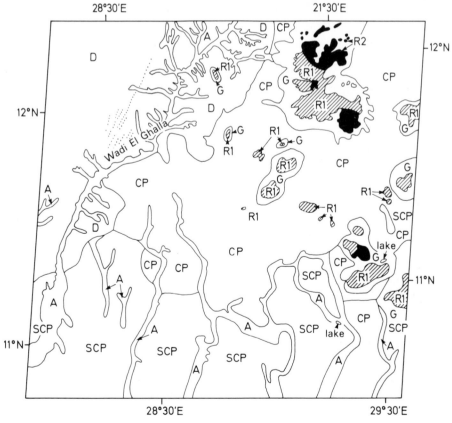

Key

R1   light coloured rock

R2   dark coloured rock*

G   gardud (the outwash aureole)

D   dune and qoz (alignments thus)

CP   clay plain

SCP   smooth clay plain

A   clay bottomed drainage

*The rock is basement complex: dark colour probably due to more ferromagnesian minerals.

Figure 8.3 Wadi El Ghalla, Nuba Mountains, Sudan: terrain units as interpreted from Landsat of 9 November 1972. Frame no. 1514. Area depicted 185 × 185 km.

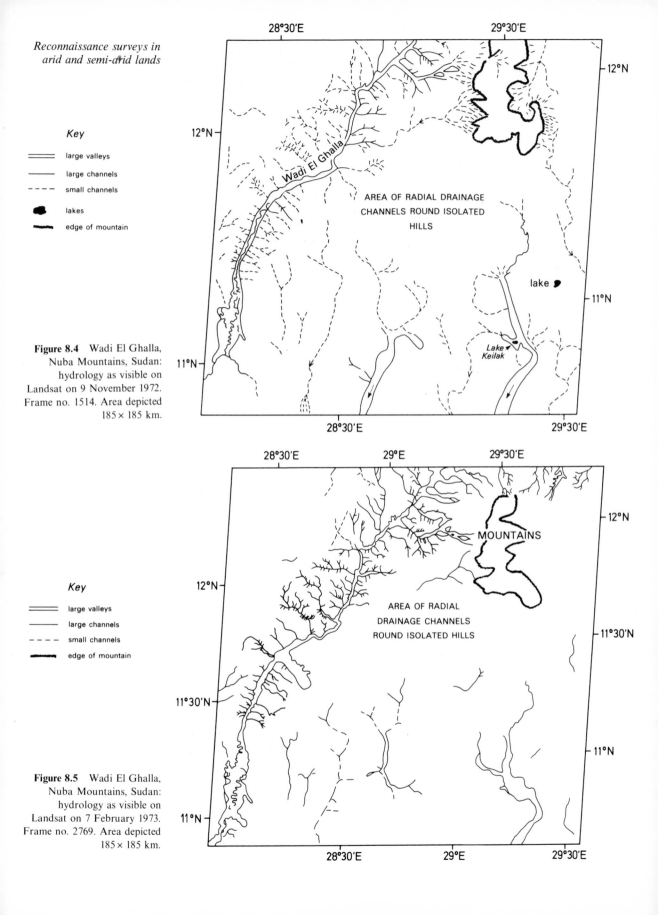

Key

large valleys
large channels
small channels
lakes
edge of mountain

Wadi El Ghalla

AREA OF RADIAL DRAINAGE
CHANNELS ROUND ISOLATED
HILLS

lake

Lake Keilak

Figure 8.4 Wadi El Ghalla, Nuba Mountains, Sudan: hydrology as visible on Landsat on 9 November 1972. Frame no. 1514. Area depicted 185 × 185 km.

Key

large valleys
large channels
small channels
edge of mountain

MOUNTAINS

AREA OF RADIAL
DRAINAGE CHANNELS
ROUND ISOLATED HILLS

Figure 8.5 Wadi El Ghalla, Nuba Mountains, Sudan: hydrology as visible on Landsat on 7 February 1973. Frame no. 2769. Area depicted 185 × 185 km.

The Sudan savanna project was a joint F A O/Sudan Government programme to develop parts of the high rainfall savanna zone of that country. The procedure was to test the value of maps of (a) soils, (b) hydrology, (c) vegetation, and (d) land use of a selected frame of varied terrain at 1:1 M with a minimum of local background information against those obtained from conventional methods of survey from low-level aerial photographs at 1:40 000 and subsequently reduced to 1:250 000. The frame chosen included El Fula, the Wadi El Ghalla and the north-western part of the Nuba Mountains, Kordofan (Fig. 8.2). The Landsat imagery permitted the identification of terrain units approximately equivalent to *complex land systems* (Christian & Stewart 1964) as a reconnaissance soils framework useful for the separation of areas of little potential value from those deserving more detailed study (Fig. 8.3). The hydrographic network was clearly interpretable especially when dry season was compared with wet season imagery (Figs 8.4, 8.5). More finger-tip channels can be identified on the latter, particularly in areas of clay soil especially as compared with sands. Little vegetation differentiation appeared which was not either a direct reflection of landform differences or of a conspicuous seasonal burning pattern (Fig. 8.6). Land use could be determined only in the crudest terms, e.g. dominantly grazing, dominantly agricultural, etc., because of the sparsity of settlement, the small size of individual land holdings, and the obscuring effect of burning. A comparison between Figures 8.7 and 8.8 shows how much this changes in 3 months. It is possible that more extensive study of colour composite imagery (not then available) might have permitted both these factors to be more closely evaluated.

The results were sufficiently favourable to extend the use of Landsat and the land system method to the resource reconnaissance of a single country at a mapping scale of 1:1 000 000. Jordan was chosen because although a considerable amount of information existed on its natural resources, it was hard to obtain a simple framework for rapidly assembling and collating it as a whole. Also, the country was well suited to synoptic landscape methods of survey because its relatively arid climate allowed cloud-free space imagery to be easily obtained and the physiography to be unvegetated and clearly visible. The dependence of the population on agriculture and pastoralism emphasized the importance of assessing the qualities of the land surface in relation to its two major development needs: control of soil erosion on the western plateau and enlargement of the irrigable area elsewhere. Both these problems are related to landforms interpretable on satellite imagery.

The *land system* method with its associated scalar hierarchy of units was used as being the most obvious way of managing the environmental information. Whereas conventional low-level aerial photography is useful for the recognition of land systems and their constituent *land facets*, the resolution and scale of Landsat imagery makes it generally more appropriate at the *land region* level (refer to Table 5.1). The availability of some ground information especially from geological maps and a field traverse, enabled terrain units approximating in size to complex land systems to be mapped, although the boundaries could not always be seen on the imagery. A block diagram was drawn for each land system, showing its constituent subdivisions. These were provisionally called *land catenas* rather than land facets because of their relatively large size and toposequential internal variations. They were sometimes visible on the imagery, an example being Wadi El Hasa (Fig. 8.11) which is visible on Figures 8.9 and 8.10. For the practical requirements of land development, it is important to recognize analogies between land catenas in different land systems. This was done by a tabulated scheme of correlation whereby recurrent land catenas, e.g. steep rock slopes, sand dunes, wadis, or mud flats, had

175

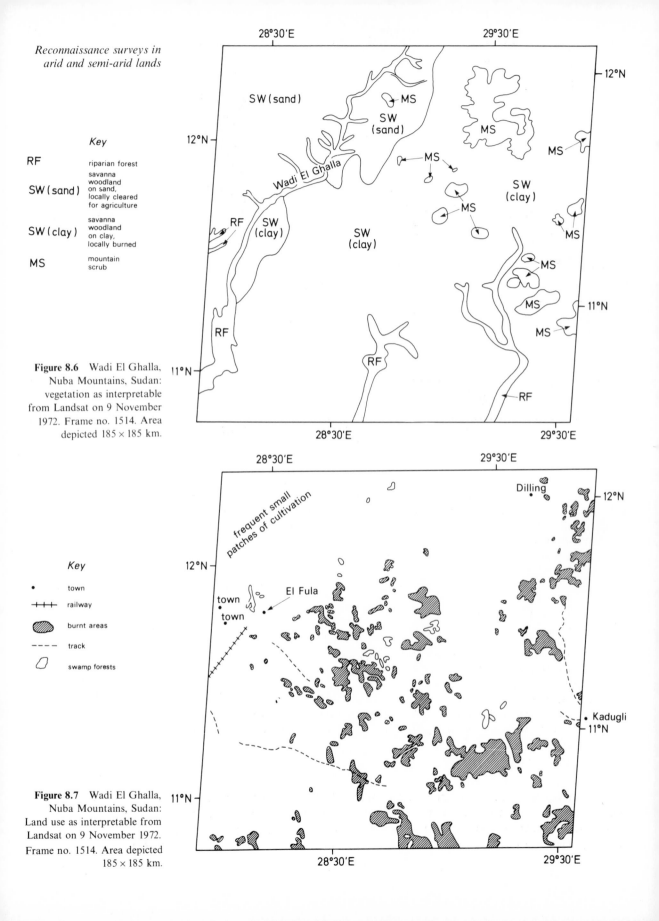

Figure 8.6 Wadi El Ghalla, Nuba Mountains, Sudan: vegetation as interpretable from Landsat on 9 November 1972. Frame no. 1514. Area depicted 185 × 185 km.

Key

RF — riparian forest

SW (sand) — savanna woodland on sand, locally cleared for agriculture

SW (clay) — savanna woodland on clay, locally burned

MS — mountain scrub

Figure 8.7 Wadi El Ghalla, Nuba Mountains, Sudan: Land use as interpretable from Landsat on 9 November 1972. Frame no. 1514. Area depicted 185 × 185 km.

Key

• town

+++ railway

burnt areas

--- track

swamp forests

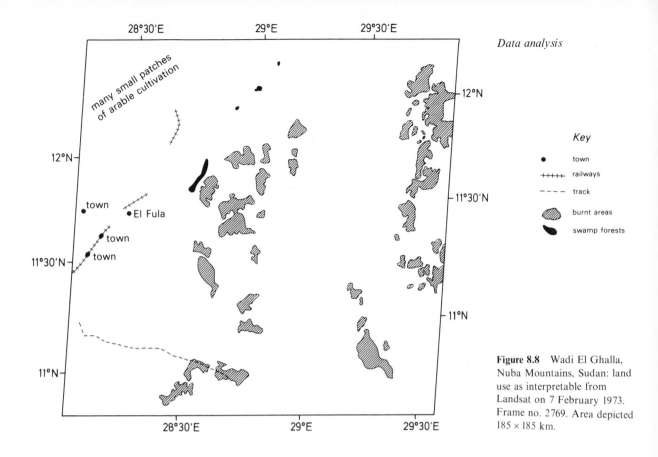

Figure 8.8 Wadi El Ghalla, Nuba Mountains, Sudan: land use as interpretable from Landsat on 7 February 1973. Frame no. 2769. Area depicted 185 × 185 km.

Table 8.3 Part of scheme of correlations between land systems and catenas in the desert zone of Jordan. The numbers separated by a stroke, e.g. 7/10 or 8/2, represent land systems. The numbers after the decimal points represent constituent land catenas. Names such as Belqa, Quweira, etc. indicate particular geological formations.

(I) *Gentle slopes (interfluves)*
   (1) Limestone: 7/10·2 (undifferentiated Belqa 1–4); 8/2·1, 8/3·1 (Belqa 5 and Dana).
   (2) Quaternary: 11/3·1
(J) *Fans*
   (1) Granite: 1/1·5
   (2) Sandstone: 2/1·4 (Quweira); 4/1·3 (Kurnub)
   (3) Quaternary: 11/3·2, 11/4·1
(K) *Wadis (incised)*
   (1) Granite: 1/1·3
   (2) Sandstone: 2/1·5 (Quweira); 3/1·7 (Ram-Um Sahm); 4/1·4 (Kurnub)
   (3) Limestone: 6/7·2 (Ajlun); 7/12·5 (Belqa 1–2); 7/8·4, 7/13·4 (Belqa 3); 17/14·4 (Belqa 4)
   (4) Basalt: 9/4·4

some analogy with each other but a closer similarity to those which were in or derived from land systems on the same rock type. A part of this is shown on Table 8.3. The Landsat imagery permitted a relatively rapid recognition of such correlations because of the distinctiveness of many of the land catenas.

177

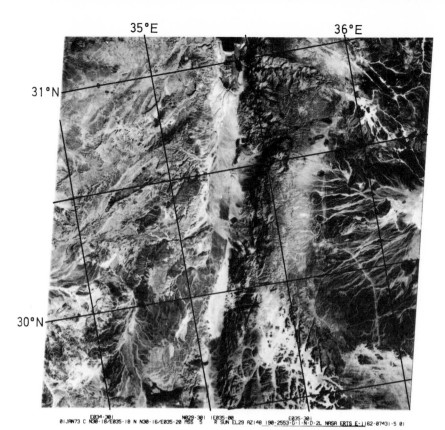

Figure 8.9 Landsat view of Palestine–Jordan frontier region showing Wadi Arabah rift valley (bottom centre) and Dead Sea (top centre).

The FAO project to produce a soil degradation map of the world at 1:5 000 000 illustrates the use of Landsat as an aid in the identification of a specific land hazard. Riquier (1978) has recognized six main types of soil degradation: water and wind erosion, salinization, and biological, chemical and physical degradation. He has suggested methods for measuring the rate at which they are occurring, and set quantitative standards for categorizing the degree of risk and the present state for any given locality. Taking water erosion as an example, standards are based on the calculation of formulae which include numerical expressions for the amount and intensity of rainfall, slope length and gradient, soil permeability, texture, structure, and organic matter content, density of vegetal cover and type and intensity of land use.

Few of these factors can be seen on Landsat, but again it is possible to use landscape units as the best guide to them at a reconnaissance scale. Such landscape units, however, have to be 'special purpose' and have to emphasize the specific attributes sought at the expense of those less useful.

The first-stage input into the 1:5 000 000 soil degradation map of the world was thus achieved by the provision of maps derived directly from the interpretation of Landsat. It would have been possible to show all the information on one map, but for practical purposes it was decided to produce two, one showing the abiotic terrain features and the other the vegetal cover and land use. Calculations of climatic and other degradational parameters were then made to characterize the units previously identified in this way. The areas covered, which included examples of all climatic zones in North Africa and the Middle East, are shown on Figure 8.12.

The units of the first map, an example of which is shown on Figure 8.13 were designated 'morphodynamic' to emphasize their relationship to the dynamic natural processes causing soil degradation and to differentiate them from

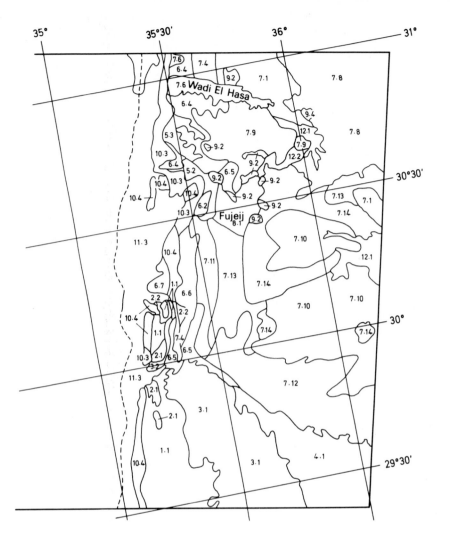

Figure 8.10 Land systems of Jordan, based on interpretation of part of Figure 8.9 and geological mapping.

Aqaba granite
1.1 hill complex

Quwevia sandstone
2.1 dissected plateaux
2.2 dissected tableland near Wadi Arabah

Ram/Um Sahm sandstone
3.1 dissected plateaux of southern desert
3.2 dissected plateau remnants near Wadi Arabah

Tabuq, Zarqa and Hathira sandstones
4.1 plateaux

Kurnub sandstone
5.2 dissected tableland remnants
5.3 major wadis draining to Jordan

Ajlun limestone
6.2 dissected tableland with drainage to Jordan
6.4 dissected area overlying Kurnub sandstone with drainage to Jordan
6.5 gullied steppe areas draining to east
6.6 Musa–Nijil wadi zone
6.7 low hills

Belqa limestone
7.1 eastern plateau slopes
7.4 dissected plateau with major wadis draining to Jordan
7.6 major gorges draining to Jordan
7.8 dissected area round Qa' El Hafira
7.9 dissected upland draining to east
7.10 surface wash fans to El Azraq and El Jafr depressions
7.11 dipslope to El Jafr
7.13 dissected dipslope to El Jafr
7.14 dissected dipslope to Sirhan, El Azraq, and El Jafr depressions

Wadi Shallala and Dana limestones
8.1 Fuj'eij plain

Tertiary basalt
9.2 plateaux in steppe zone
9.4 hamadas in arid zone

Western escarpment
10.3 Mount Nebo–Ras En Naqb section (limestone and chalk)
10.4 Ras En Naqb–Gulf of Aqaba section (sandstone and granite)

Rift valley lowland
11.3 Wadi Arabah (south of Dead Sea)

Playa systems
12.1 fine textured flats
12.2 beds of major desert wadis

179

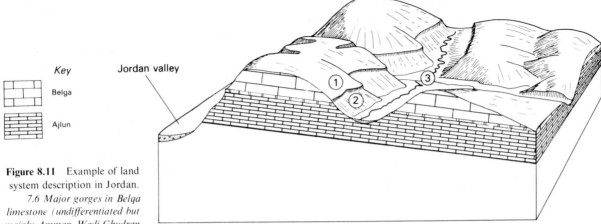

Key

Belga

Ajlun

Jordan valley

Figure 8.11 Example of land system description in Jordan.

7.6 Major gorges in Belqa limestone (undifferentiated but mainly Amman–Wadi Ghudran formation).

Climate: Steppe.

Physiography: Deep eroded gorges cut through the Belqa and into the Ajlun limestones forming high stepped escarpments. The steps are due to the outcrops of beds of greater competence, i.e. chert.

Geology: A more or less horizontally lying sequence of Amman–Wadi Ghudran cherty limestones and chalks, conformably overlying the less competent and purer Ajlun limestones.

Land Catenas:

No.	Form	Soils	Vegetation
1	erosion scarp slopes on Amman and Wadi Ghudran Formations	almost bare, stony	scattered scrub
2	erosion slopes on Ajlun limestone	almost bare, stony	scattered scrub
3	valley bottoms	stony alluvium	some irrigation

Potential for grazing and afforestation on 1 and 2, irrigation on 3.

Figure 8.12 Areas covered by Landsat interpretation for FAO soil degradation map of the world at 1:5 000 000. (1) Morocco; (2) Upper Volta, Niger, Chad, Central African Empire; (3) Lebanon, Syria, Jordan and Iraq; (4) Gambia, Guinea, Sierra Leone, etc.; (5) Iran.

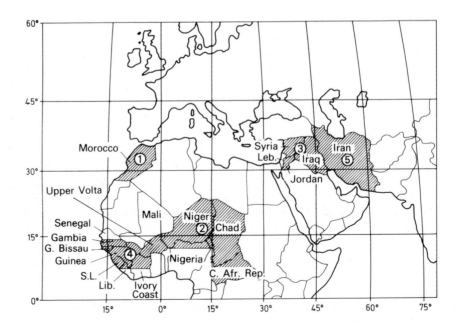

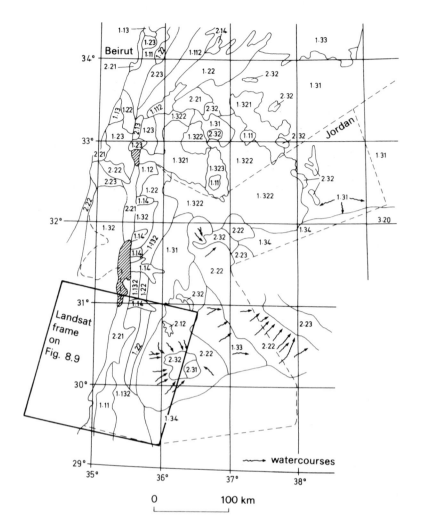

Figure 8.13 Morphodynamic units of Jordan and parts of Syria and Lebanon based on Landsat imagery, abstracted from global legend (source: Mitchell & Howard 1978).
1.11 mountains with peaks and ridges undifferentiated
1.112 ditto, but without permanent snow cover
1.132 high plateaux with much dissection
1.14 major ravines
1.22 finely dissected hills
1.23 footslopes of higher mountains
1.31 degradational plains: unchannelled
1.321 degradational plains with few channels, without rock exposures
1.322 degradational plains with few channels, with rock exposures
1.33 degradational plains: channelled
1.34 degradational plains with conspicuous aeolian action
2.21 alluvial fans and bajadas
2.22 detrital plains with braided or sinuous channels
2.23 detrital valleys
2.31 detrital plains without surface channels
2.32 seasonally flooded depressions (playas)

established landscape terminology. This was essentially a geomorphological inter- pretation which selectively showed features significant to soil degradation and omitted those irrelevant to it. The legend first separated landforms into *degradational* with *high, medium* or *low relief*, and *aggradational* types. Subdivisions of these were based on the distinction between mountainous and plateau landscapes, the texture of erosional dissection, lithology, dominant depositional or hydrological process and similar observable criteria.

Vegetal cover was expressed in percentage density classes (0–1%, 1–20%, and 20% increments thereafter) of presumed rain interception by leaves and was based on the interpretation of the depth and type of tinge on Landsat colour composite imagery taken in the dry season showing MSS4 (multispectral scanner band 4) in yellow, MSS5 in magenta, and MSS7 in cyan. This gives the effect of a 'false colour' picture. Where no red or yellow is visible, vegetation is regarded as less than 1%, a faint yellow-brown indicates 1–20%, a strong yellow or yellow-brown 20–40%, a faint brown-red 40–60%, a pronounced brown-red (xerophytic) or magenta (mesophytic 60–80%. This interpretation derived from the fact that sparser vegetation in the dry season tends to be grassy and dead and hence yellowish. With increasingly dense vegetation the amount of active photosynthesis increases due

181

largely to the increasing proportion of trees, until one obtains an almost pure continuous magenta where vegetal cover is living and complete because of the strong absorption of light in the MSS5 band and the strong reflectance in the MSS7 band.

Land use can generally be inferred from the tones and patterns on the imagery so that a distinction can be made between grazing, nucleated, non-nucleated or irrigated agricultural systems, and forestry. Figure 8.14 shows some of these distinctions.

8.5 Evaluation of results

The reconnaissance of land resources and hazards in arid and semi-arid areas is therefore best based on a landscape analysis and classification because not only does this act as a coordinating base for the main resource factors of soil, water, and vegetation, especially if done at land system and land facet scale, but also because it is exceptionally amenable to interpretation on small-scale remotely sensed imagery, notably Landsat.

The value of this approach is demonstrated by its application to three specific reconnaissance projects: the assessment of soil, water and vegetation resources and land uses in a development area of the Sudan, the provision of a correlative land system framework for Jordan, and the derivation of a first-stage input for small-scale soil degradation mapping on a worldwide basis.

8.6 Future developments

Arid and semi-arid lands will become increasingly important to world food production in the future and this will focus attention on their soil, water, and vegetal resources. Although detailed local studies will necessarily be multiplied, the need for reconnaissance assessments is likely to increase rather than decrease.

It is likely that such reconnaissance assessments will be of two types: single purpose and integrated. The former will seek to identify critical resource factors such as soil or water over wide areas, the latter to provide correlative frameworks for a variety of resources at national or international scale.

The use of imagery from orbital altitudes, especially Landsat, will increase. Its repetitive coverage, the developing sophistication of both visual and automated methods of interpretation, and the relative rapidity with which it can be made available will make it a tool of fundamental importance in many fields. The improved resolution now available from Landsat 3 and the wider spectral coverage to be available from Landsat D point the way to wider applications in the future, extending to the replacement of conventional aerial photography for some purposes.

Acknowledgement

Thanks are due to Dr John Howard and the FAO Remote Sensing Unit under whom the project work reported here was done.

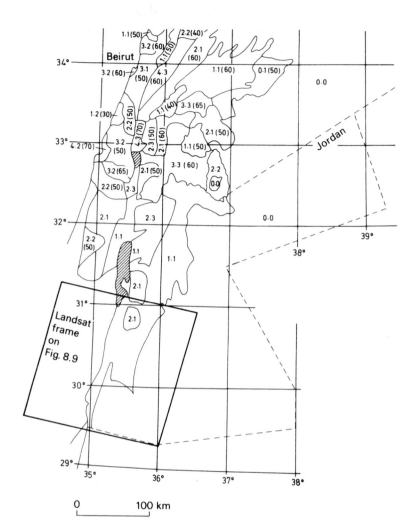

Figure 8.14 Vegetal cover and land use of Jordan and parts of Syria and Lebanon, based on Landsat imagery.

symbol	Vegetal cover (%)	symbol	Land use type
0	0–1	0	none visible
1	1–20	1	grazing or grass fallow
2	20–40	2	nucleated agriculture
3	40–60	3	non-nucleated agriculture
4	60–80	4	irrigation
5	80–100	5	forestry

The symbols are given thus: 3.4(50) indicates first the vegetal cover, then the land use type and then the areal percentage of the unit it covers if this is less than 100%.

References

Christian, C. S. and G. A. Stewart 1964. Methodology of integrated surveys. *Unesco Conf. on Principles and Methods of Integrating Aerial Survey Studies of Natural Resources for Potential Development, Toulouse,* 233–80, Unesco.

Hills, E. S. (ed.) 1966. *Arid lands: a geographical appraisal.* London: Methuen.

Joly, F. 1957. Les milieus arides. Définition, extension. *Notes Marocaines, Rabat* **8**, 15–30.

Meigs, P. 1953. World distribution of arid and semi-arid homoclimates. *Rev. Res. on Arid Zone Hydrology,* Unesco NS/A7/37 of 21/8/1961 and Maps UN 392 and 393.

Mitchell, C. W. 1975. The application of Landsat 1 imagery to the Sudan Savanna Project. *J. Br. Interplanetary Soc.* **28**, 659–72.

Mitchell, C. W. 1978. The use of Landsat imagery in a land system classification of Jordan. *J. Br. Interplanetary Soc.* **31**, 283–92.

Mitchell, C. W. and J. A. Howard (compilers) 1978. *The application of Landsat imagery to soil degradation mapping at 1:5 000 000.* Rome: FAO.

Mitchell, C. W., R. Webster, P. H. T. Beckett and B. Clifford 1979. An analysis of terrain classification for long range prediction of conditions in deserts. *Geogl J.* **145**, 72–85.

Riquier, J. 1978. *A methodology for assessing soil degradation in FAO/UNEP.* Report on the FAO/UNEP expert consultation on methodology for assessing soil degradation, World Assessment of Soil Degradation, proj. no. 1106–75–05, Rome, 25–59.

9 Integrated resource survey as an aid to soil survey in a tropical flood plain environment

Alison Cook

The feasibility of using an integrated survey approach as a basis for obtaining soil survey information is examined in this chapter with reference to a case study of the Lower Rufiji Basin in Tanzania. In order to put this study into perspective, the information requirements of a soil survey and possible survey approaches will first be considered.

9.1 Information requirements for soil survey

The information needs for soil survey depend on its declared aims, these being generally viewed as much wider than simply the mapping of soil units (Trudgill & Briggs 1977). Indeed Young (1973) includes as essential components the definition of alternative land uses, estimation of crop yields and assessment of soil response to changing land uses. The aim of a soil survey is also viewed in terms of the intended user. One dichotomy is between the broad general-purpose and the narrow special-purpose survey. The British Soil Survey which is based on the former approach (Avery 1973) has failed to be of great practical use either to planners or agriculturalists (Trudgill & Briggs 1977). However, a special-purpose survey is also fraught with difficulties in that the knowledge required to select the soil properties appropriate to the purpose may not be to hand (Gibbons 1961). For example, crop –soil relationships are often insufficiently understood, particularly in developing countries, to permit estimation of crop yields from soil survey information (Avery 1962). At best, recommendations of alternative land uses can be made. There is often no option, therefore, but to conduct a general-purpose survey (Mulcahy & Humphries 1967), as was the case in the present study.

Judgement of the usefulness of information derived from a soil survey, has been viewed as relating to the homogeneity of mapping units (Bie & Beckett 1971). Soil being notoriously variable spatially, all mapping units are likely to include local impurities (Beckett & Webster 1971, Webster & Wona 1969). Furthermore, it is likely that widely defined units will contain a greater degree of unit heterogeneity and thus be less predictable (Bie & Beckett 1971, Mitchell 1973).

9.2 Possible survey approaches

Possible survey methods fall into two fundamentally different approaches: first, the conventional soil survey in which units are defined and mapped purely on the basis of soil profile characteristics; secondly, the integrated survey in which the terrain is classified into units by the assessment of the total environment including its relief, drainage, geology, vegetation and soils. The terrain units can subsequently be related specifically to soil properties.

One theoretically-based argument in favour of the integrated approach states that by the inclusion of all environmental properties, a soil survey is more likely to meet the varied aims of a general-purpose survey by maximizing the information input (Gibbons 1961). Practically-based arguments relate to the heavy reliance placed on the interpretation of aerial photography and other remotely sensed imagery which provide the required integrated view of the terrain. Since this results in both reduction of field time and personnel numbers, it is of great importance particularly in developing nations where photo-interpretation is also appropriate to the large and often inaccessible areas. The readily and cheaply available Landsat imagery has already been applied to integrated survey work in Africa (King & Blair-Rains 1974, Kirkula *et al.* 1978, Mitchell 1975, Parry 1974, 1978). Cost-effectiveness should also follow as a consequence of time and staff reductions.

The most widely applied integrated survey for soil data collection has been the land system approach of Christian and Stewart and the M E X E group (Ch. 5) which has been applied in Australia, Papua-New Guinea and many African countries. In Tanzania, integrated surveys have been carried out by Baker (1970) for the whole country; Cook (1974, 1976, 1977) for the Rufiji, Wami and Mtera Basins; Kirkula *et al.* (1978) for the Rukwa Region; and Strömquist (1976) for the Great Ruaha Basin. Variations in detail of approach do occur according to the nature of terrain: the more complex the terrain, the more classification levels are required (Lawrence *et al.* 1977); the strength of correlation between surface features and soil properties crucial to most kinds of soil survey also varies with terrain type as does the ground sampling framework in terms of arrangement and density of sample points (Ch. 3).

Have the applications of the land system approach shown its superiority over conventional surveys, in the quality and usefulness of the soil data collected? Certainly land capability estimates have been stressed, although these are generally of a qualitative nature.

As an example of unit homogeneity comparisons Perrin & Mitchell (1969), reported a 67·6% coefficient of variation for per cent clay in land units drawn from arid regions, although in the Oxford area 25·6% variation was found for this factor (Webster & Beckett 1968). The latter compares favourably with an average coefficient of variation of 28·3% found for mapping units of soil surveys carried out in many different parts of the world (Beckett & Webster 1971). These results suggest that land unit homogeneity is at least as good as that for conventional soil mapping units. However, more quantitative or strictly comparable data are required to test this assumption fully.

9.3 The case study: aims and location

To test the feasibility of the integrated approach to soil survey in this case study, particular attention was paid to the quality of the soils information derived from the terrain units, in terms of purity and homogeneity, and hence the level of land use planning decision which could be reliably based on this information.

The environment selected for this feasibility study was a flood plain and the adjoining northern river terrace (1080 km² in area), the former chosen because its complexity was thought to provide a good test of the approach. Wright (1973) for instance has commented on the 'intricate mosaic of morphological sites' on the Fitzroy flood plains, Australia. In Tanzania, such areas pose some difficult land use

185

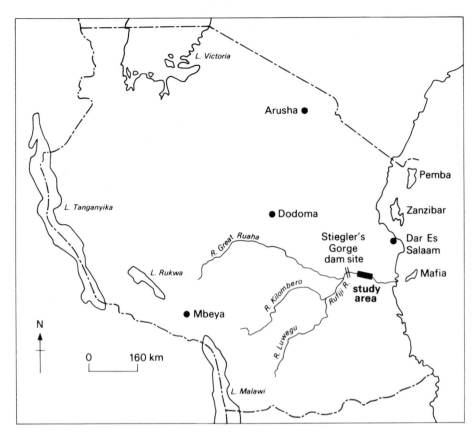

Figure 9.1   Location of the
study area.

problems in that although of obviously high agricultural potential they are also
hazardous and in fact do not support high population densities. River terrace areas,
on the other hand, represent the environment of highly leached soils and rainfed
agriculture which covers the greater part of Tanzania, and upon which the majority
of the population relies for its livelihood (Cook 1974).

The survey area chosen included part of the Lower Rufiji Basin (Fig. 9.1). The
flood plain comprises a mosaic of abandoned river channels, levees and shallow
depressions supporting respectively a sparse shrub vegetation, intensive cultivation
or woodland, scattered cultivation or tall grassland. Amplitude of relief is minimal
(4 m (12 ft) maximum) whilst altitude varies from 17 m (50 ft) to 50 m (150 ft). The
annual flood occurs during March to June though intensities and spatial extent vary
greatly. The northern terrace situated 17 m (50 ft) above the flood plain (Fig. 9.2)
again has very level relief. Soils are mainly sands supporting a deciduous woodland,
with clays in extensive shallow depressions underlying tall grassland. Rainfall varies
from 700 mm in the east to 893 mm in the west (Cook 1974).

A land use problem specific to the area is the possible construction of a power
station and dam at Stiegler's Gorge (Fig. 9.1). This would greatly alter the flooding
regime upon which the present agricultural system relies. Likely change of soil
properties under mechanized irrigation is therefore a relevant question in this area.
A second problem resulted from the movement of the local population for reasons
of safety from the flood plain to new villages on the river terrace (Fig. 9.2). The
terrace soils are unmapped, so their appraisal in terms of the most suitable crops is a
pressing issue.

186

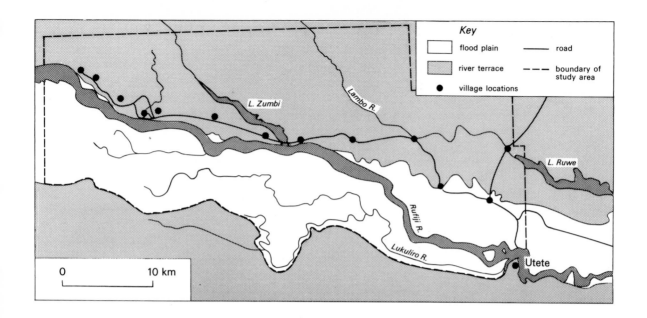

Figure 9.2 The study area.

A previous soil survey had been carried out on the flood plain by conventional techniques to assess future irrigation possibilities (Anderson 1961). This therefore provided a comparison with the present integrated approach.

9.4 Experimental procedures

A preliminary interpretation was carried out in order to classify the terrain of the study area. The available photography was of the scales 1:25 000 (June 1964) and 1:48 000 (June 1966). The latter had also been used to compile very good quality 1:50 000 semi-controlled mosaics, equivalent in scale and area to the contour maps. No detailed geological map was available. Working first with the mosaics, land system and land unit boundaries were delimited on the basis of tonal and textural characteristics. Proceeding from small to large scale, stereoscopic inspection of the contact prints was then carried out. Each land unit was described in terms of its relief, drainage and vegetation as well as a prediction of the underlying soil textural type. Criteria for the photo-identification of the vegetation types is given in Table 9.1.

A programme of ground control was devised for testing the overall validity of the terrain classification and homogeneity of the land units with respect to their physical and chemical soil properties. A reconnaissance visit to the area indicated that location of sample sites would have to be on a subjective basis, due to the problems of inaccessibility and location. The sampling framework was therefore based on road and river access points, whilst also attempting to achieve an even distribution amongst the delineated land units.

Data recorded at each sample site comprised a general description of location, topography and drainage. Vegetation was recorded as plant species and per cent cover for an area of 25 m² or as type of land use. A full soil profile description was made to a depth of 100 cm. On the basis of these observations, each site was given a land unit classification. The vegetation was classified into broad physiognomic

187

Table 9.1 Criteria for photo and field identification of vegetation types.

Vegetation type	Photo-identification	Field identification
(1) Riverine forest	Black to dark grey tone, fairly smooth texture. Tall trees with merging crowns	A closed and layered canopy. Tree height of 10–50 m. Bordering water courses
(2) Dense woodland	Dark grey tone, fairly smooth texture. Trees of medium height with merging crowns	Closed canopy of single storey
(3) Open woodland	Mid grey tone, mottled texture. Trees of medium height. Individual crowns distinguishable and ground layer visible	Open canopy of single storey. Ground cover $> 50\%$
(4) Wooded grassland	Mid grey tone, smooth texture with pattern of dark grey tree crowns of either scattered or clumped distribution	Tree canopy cover $< 50\%$, $> 10\%$
(5) Grassland	Mid to light grey tone, smooth texture. Tree crowns of very low density	Tree canopy cover $< 10\%$. Grass cover $> 50\%$
(6) Sparse grassland	Mid to light grey tone, mottled texture. Evidence of much bare ground	Tree canopy cover $< 10\%$. Grass cover $< 50\%$
(7) Thicket	Mid grey tone, fairly smooth texture. Heights of plants not discernible at scales used. No ground layer visible	Dense shrub community of < 3 m height

groupings using nomenclature and field definitions adapted from Gillman (1949) and listed in Table 9.1. The soil profiles were classified on the basis of dominant colour and texture and used as one means of assessing land unit homogeneity.

Soil samples were taken from 15, 30 and 60 cm depths, a procedure adopted in preference to sampling each genetic horizon so that the soil profiles could be more readily compared in terms of their crop growth potential and hence land capability. The following routine chemical analyses were carried out: pH (1: 2·5 soil:water ratio); per cent organic carbon (C); per cent total nitrogen (N); available phosphorus (P); exchangeable bases-calcium (Ca), magnesium (Mg), potassium (K), sodium (Na) and manganese (Mn); exchange acidity (H). From these total exchangeable bases (TEB), cation exchange capacity (CEC) and per cent base saturation were calculated. The exclusion of physical soil factors was due to lack of laboratory facilities. Examination of the chemical data, particularly with reference to crop growth requirements, was used as a second means of evaluating land unit homogeneity.

9.5 Evaluation of the results

9.5.1 *The land system classification*

Two land systems, the flood plain and river terrace, were delimited on the mosaics. Boundary recognition was straightforward due to the sharp contrast between the depositional patterns of the flood plain and the generally wooded appearance of the adjoining terrace (Fig. 9.3).

Figure 9.3 Flood plain land units: (1) former river channel; (2) levee; (3) depression; (4) levee over former river channel; (5) depression over former river channel; (6) depression over levee.

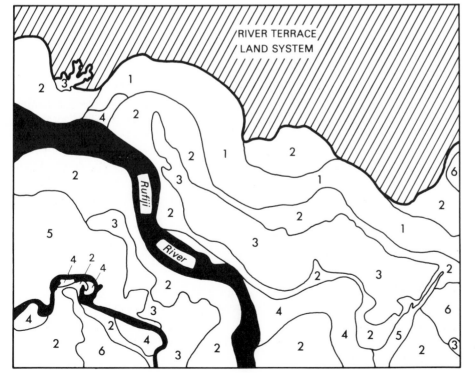

189

Recognition of the five flood plain land units was achieved using the criteria of shape and form of depositional pattern and secondarily, by use of vegetation and land use characteristics (Table 9.2). The former river channel land unit had a diagnostic shape, downcut form and sparse grass cover with visible indication of the underlying soil type (Fig. 9.3). The levee unit was less accurately delimited in that no relief form was discernible. Recognition was based on land use intensity as shown by the high density of light-toned fields recently prepared for crop planting (Fig. 9.4). The soil type was assumed to be fine sand or loam, since flood waters deposit medium-textured material on the river banks. The levee grades into the depression unit characterized by sparse cultivation and a darker-toned grassland, indicative of less well drained clay soils resulting from deposition of fine-textured alluvium (Fig. 9.4). Two 'composite' units were also recognized both having the shape of a former channel but the surface features of cultivation in one case, and tall grassland in the other, suggested an overlying medium-textured (levee) and fine-textured (depression) deposit respectively (Fig. 9.3).

Depositional pattern was therefore the chief means of recognizing the flood plain land units.

By contrast, the criterion used for delineating the six river terrace land units was predominantly vegetation, specifically the degree of canopy closure, height and distribution of trees (Table 9.2). Thus a distinction was made between open and dense woodland land units and two types of wooded grassland with scattered and clumped tree distribution in the shallow depressions, all of which had boundaries of a gradational nature (Fig. 9.5). Prediction of soil type was based on tree density, the woodland vegetation being taken to indicate a freely drained sandy soil, whereas the wooded grassland suggests poorly drained clay soil with restricted rooting depth. The clumped tree distribution is associated with old termitaria providing a greater rooting depth in otherwise poorly drained, probably saline clays. For the two major river valley land units, riverine forest (river bank sand) and grassland (clay), topography was a useful criterion of recognition, in addition to vegetation (Fig. 9.5).

Testing the overall accuracy of the land system classification was achieved by comparing the preliminary photo-interpretation of the land units with those identified in the field for 22 sample sites on the flood plain and 36 on the river terrace. In the case of the flood plain sites, four errors were revealed. Two additional 'composite' land units were recognized – depression over levee and former river channel over depression – which accounted for two of these. Of the 36 river terrace sample sites, errors were made in two cases due to the occurrence of thicket vegetation (predicted as dense woodland), which was then classified as an additional land unit. These results show overall accuracy to be high, particularly in the case of the river terrace. This is especially so in that the error due to the unidentified thicket could be rectified after establishing ground–photo correlations, although this did not prove to be the case on re-examining the erroneous flood plain sample sites.

9.5.2 *Land unit homogeneity*

High overall classification accuracy is only a useful basis for soil survey if the identified land units are homogeneous in terms of their soil properties. Considering first the physical soil properties as revealed by the field profile descriptions, the flood plain land units differed in the degree of variability shown, although the general soil type was as predicted (Tables 9.2 and 9.3). The former river channel unit showed the

Table 9.2 Photo characteristics of the land units.

Land system and land units	Relief and drainage	Vegetation/land use	Predicted soil type
Flood plain			All soils seasonally poorly drained
(1) Former river channel	Braided network of active channels – level	Sparse grassland. Much bare ground	Coarse sand
(2) Levee	A few tributary streams – level or gently sloping	Intensive cultivation or open woodland	Fine sand or loam
(3) Depression	Dendritic network of streams – level	Scattered cultivation and wooded grassland	Clay
(4) Levee over former river channel	A few active channels – level	Intensive cultivation	Fine sand or loam over coarse sand
(5) Depression over former river channel	A few active channels – level	Scattered cultivation and wooded grassland	Clay over coarse sand
River terrace			
(1) Terrace surface A	A few annual streams – level or gently sloping	Open woodland with very scattered cultivation	Freely drained sand
(2) Terrace surface B	Annual streams more numerous than A – level or gently sloping	Dense woodland – canopy closed and trees smaller than in A	Freely drained sand
(3) Depression A	Numerous annual streams – level	Wooded grassland with scattered trees	Clay – seasonally poorly drained
(4) Depression B	Numerous annual streams – level	Wooded grassland – clumped trees	Saline clay – seasonally poorly drained
(5) Major river valley A	Perennial stream – level	Riverine forest	Coarse sand – seasonally poorly drained
(6) Major river valley B	Annual tributary streams – level	Grassland	Clay – seasonally poorly drained

191

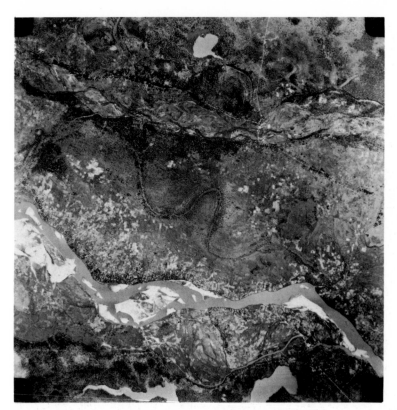

Integrated survey of a tropical flood plain

Figure 9.4 Flood plain land units: (1) former river channel; (2) levee; (3) depression; (4) levee over former river channel; (5) depression over former river channel; (6) depression over levee; (7) former river channel over depression.

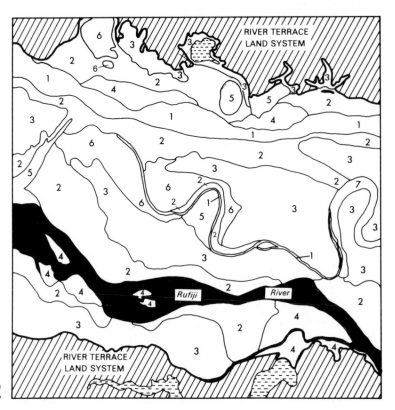

192

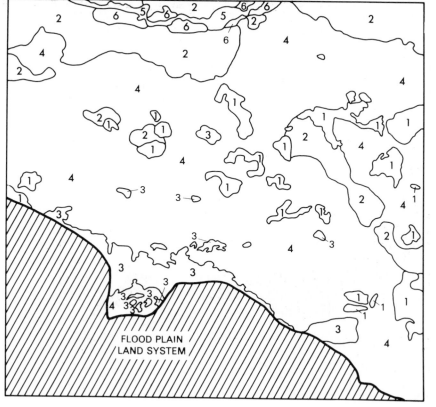

Figure 9.5 River terrace land units: (1) terrace surface, dense woodland; (2) terrace surface, open woodland; (3) terrace surface, thicket; (4) depression, wooded grassland with scattered trees; (5) major river valley, riverine forest; (6) major river valley, grassland.

193

Table 9.3 Relationship between the interpreted land units and the observed soil type.

Land system/land unit (no. of sample sites)	Observed soil type (no. of profiles)
Flood plain	
Former river channel (3)	Brown sand
Levee (8)	Brown loam
Depression (5)	Brown clay
Levee over former river channel (3)	Brown loam over sand
Depression over levee (2)	Brown clay over loam
Former river channel over depression (1)	Brown sand over clay
River terrace	
Terrace surface – open woodland (19)	(1) Red sand (14)
	(2) Brown sand (5)
Terrace surface – dense woodland (2)	Red sand
Terrace surface – thicket (3)	(1) Red sand
	(2) Brown sand
Depression – wooded grassland with scattered trees (11)	(1) Black clay (10)
	(2) Brown sand with induration (1)
Major river valley – riverine forest (1)	Brown sand
Major river valley – grassland (1)	Black clay

greatest uniformity being consistently a pale brown, structureless, excessively drained coarse sand. The darker, heavier textured, buried organic horizons did show variation in terms of number and position. The soils of the levee land unit were most commonly fine sand or sandy loam, usually exhibiting three or four distinct textural horizons within the profile examined. Mottling due to ferric oxides occurred throughout these soils and some moisture retention was evident in the dry season in contrast with the former river channel soils. The soils of the depression land unit were also characterized by narrow textural horizons usually of clay loam and clay texture with occasional loamy sands such that structure was generally strong and blocky. Ferric oxide mottling occurred at all depths and subsoils were moist although the topsoil was dried out and exhibited vertical cracking. The three 'composite' units showed the expected properties of the component layers comprising them. The depression over former river channel unit was not examined in the field. Although all except one of the flood plain land units had soils with a fair degree of heterogeneity, particularly with respect to texture, as would be expected in an alluvial environment, it is possible to generalize about their physical properties such that comparative statements can be made about their land use potential (see following section).

By comparison, the soils of the river terrace land units showed greater internal uniformity. The three terrace surface land units all contained excessively drained, poorly structured coarse sands with some degree of colour variation from red to brown (Table 9.3). The dense woodland and thicket land units differed in having a better developed organic horizon and some moisture at depth since they occurred near large lakes or perennial streams. Ten of the depression land unit sites were occupied by a fairly uniform black clay soil, with strong structure showing surface cracking (Table 9.3). Greyish mottling of ferrous oxides indicated a less well aerated subsoil which was generally moist and frequently contained calcium carbonate nodules. At one depression site a brown sand with indurated subsoil was found. One site was visited in each of the major valley land units sufficient only to confirm the predicted soil type of a coarse sand under the riverine forest and a black clay under the grassland. The two remaining land units, depression with clumped trees and the escarpment, were not visited in the field.

A further indication of the degree of variability is given by the results of the soil chemical analyses (Table 9.4), quoted as means (except where only one sample site

Table 9.4 Mean values and standard deviations for selected soil chemical analyses.

Land units and soil type (no. of profiles)	Depth (cm)	pH	C(%)	N(%)	P (ppm)	TEB	H	CEC	% base satura-tion
						(meq/100g soil)			
Flood plain									
Former river channel	15	6·9	0·32	0·046	70	15·79	2·36	18·14	82
Brown sand (3)		(0·6)	(0·22)	(0·016)	(28)	(16·88)	(1·25)	(17·30)	(13·49)
	30	7·0	0·13	0·048	47	5·98	2·61	8·59	69
	60	7·2	0·13	0·025	46	8·99	2·72	11·71	68
Levee	15	7·1	0·62	0·095	227	37·40	2·65	40·05	92
Brown loam (8)		(0·4)	(0·29)	(0·054)	(75)	(19·57)	(0·43)	(21·53)	(4·24)
	30	7·0	0·36	0·052	161	24·13	2·75	26·88	87
	60	7·1	0·57	0·058	211	26·27	2·99	29·26	85
Depression	15	6·2	0·95	0·125	205	44·04	3·43	47·47	92
Brown clay (5)		(0·8)	(0·46)	(0·100)	(40)	(19·11)	(1·13)	(19·61)	(3·63)
	30	6·4	0·63	0·081	95	43·14	3·34	46·48	93
	60	6·6	0·61	0·091	68	33·53	3·34	36·87	91
Levee over former	15	6·5	0·77	0·086	123	36·36	3·29	39·65	86
river channel		(0·6)	(0·29)	(0·026)	(50)	(23·47)	(0·75)	(23·39)	(12·50)
Brown loam over sand (3)	30	7·1	0·16	0·032	241	29·29	2·44	31·73	90
	60	7·6	0·12	0·026	101	5·10	2·54	7·60	66
Depression over levee	15	6·1	0·76	0·106	81	47·98	3·65	51·13	94
Brown clay over loam (2)	30	6·2	0·42	0·043	35	38·39	3·91	42·30	91
	60	6·8	0·15	0·030	110	18·92	3·28	22·20	84
Former river channel	15	7·3	0·05	0·025	142	6·84	3·00	9·84	70
over depression	30	7·6	0·03	0·020	113	6·39	2·33	8·72	73
Brown sand over clay (1)★	60	6·2	0·91	0·115	55	56·76	2·73	59·49	95
River terrace									
Terrace surface (open	15	7·0	0·70	0·068	45	6·89	1·27	8·16	82
woodland)		(0·4)	(0·31)	(0·027)	(24)	(4·59)	(0·84)	(5·14)	(8·47)
Red sand (6)	30	7·1	0·42	0·048	38	7·62	1·75	6·13	75
	60	6·6	0·29	0·044	38	8·35	2·36	10·71	77
Brown sand (3)	15	6·5	0·54	0·086	20	8·35	1·11	9·46	84
		(0·5)	(0·29)	(0·021)	(9)	(4·40)	(0·75)	(3·68)	(14·47)
	30	6·4	0·77	0·041	13	5·02	0·95	5·97	84
	60	6·4	0·58	0·027	14	5·40	0·78	6·18	88
Terrace surface (open	15	7·1	0·82	0·079	54	6·99	1·69	8·68	70
woodland now cultivated)		(0·3)	(0·18)	(0·013)	(44)	(1·91)	(0·53)	(1·71)	(16·55)
Red sand (7)	30	7·2	0·46	0·053	36	5·73	1·83	7·56	66
	60	6·7	0·17	0·032	31	3·20	2·26	5·46	63

Table 9.4 – *continued*.

Land units and soil type (no. of profiles)	Depth (cm)	pH	C(%)	N(%)	P (ppm)	TEB	H	CEC	% base satura-tion
							(meq/100g soil)		
Brown sand (2)	15	6·4	0·93	0·095	9	7·69	0·92	8·61	86
	30	6·9	0·40	0·051	5	7·92	0·92	8·84	82
	60	6·6	0·19	0·027	18	7·76	1·00	8·76	85
Terrace surface (dense woodland) Red sand (2)	15	7·0	1·16	0·155	113	9·25	2·25	11·50	84
	30	7·0	0·66	0·075	265	5·22	2·09	7·31	71
	60	7·0	0·29	0·033	115	3·40	2·34	5·74	56
Terrace surface (thicket) Red sand (1)★	15	6·5	0·62	0·107	70	4·38	2·50	6·88	64
	30	6·3	0·33	0·057	49	7·88	3·00	10·88	72
	60	6·4	0·17	0·067	53	6·41	2·17	8·58	75
Brown sand (2)	15	6·9	1·12	0·075	44	4·65	1·00	5·65	81
	30	7·0	0·72	0·039	32	8·14	1·66	9·80	79
	60	7·0	0·45	0·031	69	10·95	1·42	12·37	89
Depression (wooded grassland with scattered trees) Black clay (10)	15	6·6	0·90	0·103	76	29·00	2·47	31·47	92
		(0·5)	(0·17)	(0·059)	(35)	(14·26)	(1·21)	(14·63)	(5·67)
	30	7·2	0·62	0·073	33	30·43	1·58	32·01	94
	60	7·7	0·37	0·050	27	29·42	1·36	30·78	93
Brown sand with induration (1)★	15	7·2	0·31	0·103	2	8·51	3·33	10·84	79
	30	7·2	0·24	0·045	2	14·20	2·73	16·9	84
	60	—	—	—	—	—	—	—	—
Major river valley (riverine forest) Brown sand (1)★	15	5·5	1·34	0·107	7	2·75	1·67	4·42	62
	30	5·3	0·84	0·040	8	2·41	1·67	4·08	59
	60	6·2	0·57	0·017	7	2·84	1·67	4·51	63
Major river valley (grassland) Black clay (1)★	15	7·7	0·53	0·117	—	—	—	—	—
	30	7·8	0·47	0·069	19	16·63	0·00	16·63	100
	60	7·0	0·27	0·026	8	10·62	0·33	10·95	97

★Individual values quoted. Standard deviations in brackets.

was visited). Standard deviations have been included for the 15 cm depth for soil types with three or more sample sites. The soil variants within land units have been tabulated separately, as have cultivated sites, since it is useful to isolate variation due to this factor. For the flood plain land units the mean values indicate generally neutral pH values, fairly high C, N and P levels and total bases. These are lowest for the former river channel soils, being considerably higher for the levee soils and highest in the depression soils. In the 'composite units', the different depositional layers resemble the soil profile of similar composition (the 15 and 30 cm depths belonging to the overlying deposit in each case). Although the mean values point to chemical differences between the flood plain land units, the standard deviations indicate considerable variability within units and overlap between units.

The general chemical nature of the soils of the terrace surface land units differ markedly from those of the flood plain having considerably lower nutrient levels. Inter-unit differences are not marked except for somewhat higher C, N and P levels under dense woodland and thicket. The cultivated sites, however, are not consistently different from the undisturbed areas. The distinction made between red and brown sands is reflected only in the slightly lower pH and P levels of the latter. The standard deviations again indicate that even with the subdivision of the land units into soil types variability is still great. In contrast to the sands, the black clay of the depression land unit exhibits high base levels and base saturation. The one brown sand profile within this land unit has much lower base levels and per cent saturation values, indicating unit variability, as do the high standard deviations for the black clay soil type. The major river valley land units show marked contrast of soil characteristics but unit variability was not tested in this case.

An examination of the physical and chemical soil properties has therefore shown up differences in the degree of land unit variability, the greatest heterogeneity being exhibited by the chemical characteristics and by the flood plain units.

9.5.3 *Soil variability and land use potential*

Despite the intra-unit variability encountered on examining the soil data, a valid test of the usefulness of the information can only be carried out with reference to land use potential, in that it may in fact be possible to make certain land use recommendations for land units exhibiting some measure of heterogeneity.

In order to make this assessment the chemical properties of pH, N, P and K at the 15 and 30 cm depths have been expressed for each land unit/soil type as a range of values (Table 9.5). It was found that detailed information on crop–soil relationships in Tanzania was very difficult to obtain and in fact, a general guideline only has been taken here in order to assess the heterogeneity of the land units with regard to their potential for crop growth. Satisfactory levels (i.e. no fertilizer required) are assumed as 0·1% N, 25 ppm P and 0·25 meq/100 g soil K. Low values are 0·05 – 0·08% N, 10–15 ppm P and 0–0·125 meq K. The assessment of the land units' physical suitability for crop growth has been based on the field profile descriptions and site characteristics. These data have been summarized in Table 9.6 which shows the physical limitations and nutrient status of each land unit, the fertilizer requirement, crop recommendations and land use potential class. The classification used was a primary division into the four following categories:

(1) wide choice of crops; high yields; no limitations;
(2) some limitations and more restricted crop range; high to moderate yields;
(3) more serious limitations; moderate to low yields;
(4) land marginal for crop growth, its use depending on economic conditions; low yields.

Limitation suffixes have also been used, devised to be appropriate to the main physical and chemical limitations of the soils in the study area (Table 9.6). The possible effects of flood control or continued cropping on soil fertility have not been considered, the assessments being related to the soil conditions at the time of survey.

From the viewpoint of physical soil and site characteristics, the main limitation of the flood plain land units is the flooding itself with waterlogging a problem for the depressions and lower water holding capacity for the former river channels. These differences are reflected in the wide range of crops which can be grown on the levee and the more restricted range in the depression, whilst cultivation is uneconomic on the former channels. The river terrace land units are more limited physically, the sandy soils storing little water so that they can support only drought tolerant crops whereas the depression clay soils suffer from seasonal flooding and poor aeration

Table 9.5 Range of values for selected chemical soil properties.

Soil property	pH	pH	N%	N%	P(ppm)	P(ppm)	K (meq/100g soil)	K (meq/100g soil)
Land unit/soil type Sample depth	15 cm	30 cm	15 cm	30 cm	15 cm	30 cm	15 cm	30 cm
Flood plain								
Former river channel	6·3–7·5	6·7–7·3	0·030–0·062	0·028–0·068	48– 98	38– 54	0·06–0·14	0·05–0·09
Levee	6·7–7·4	6·4–7·6	0·041–0·149	0·028–0·076	158–396	122–200	0·11–0·91	0·11–0·31
Depression	5·4–7·0	6·2–6·6	0·115–0·135	0·051–0·111	169–241	37–153	0·11–1·15	0·39–1·97
Levee over former river channel	5·8–7·2	4·8–7·2	0·082–0·090	0·030–0·034	64–186	120–230	0·10–0·48	0·16–0·28
River terrace								
Terrace surface (open woodland)								
(1) Red sand	6·6–7·4	6·8–7·4	0·041–0·095	0·040–0·056	21– 69	19– 57	0·12–0·46	0·16–0·50
(2) Brown sand	6·1–7·2	6·1–6·8	0·067–0·115	0·032–0·046	7– 27	3– 20	0·18–0·45	0·09–0·49
Terrace surface (dense woodland and thicket)								
(1) Red sand	6·5–7·1	6·3–7·3	0·043–0·281	0·040–0·098	64–134	12–368	0·15–0·43	0·22–0·36
(2) Brown sand	6·6–7·2	6·5–7·5	0·073–0·127	0·036–0·072	41– 47	28– 35	0·24–0·38	0·15–0·28
Depression (wooded grassland with scattered trees)								
Black clay	6·1–7·1	6·3–8·1	0·044–0·162	0·053–0·093	41–111	11– 55	0·26–0·72	0·21–0·71

Table 9.6 Land use potential of the land units*.

Land unit and soil type	Physical limitations	Nutrient status	Fertilizer requirement	Crops recommended	Land use potential class
Flood plain					
Levee. Brown loam	Flooding, otherwise no physical limitation	N, P, K sufficient; pH satisfactory	No fertilizers required	A very wide range – rice, maize, cotton, fruit trees, millet, cassava, sugar cane, tobacco, cashew	High – 2 f
Depression over levee. Brown clay over loam	Flooding; otherwise no physical limitation	N, P, K sufficient, pH satisfactory	No fertilizers required	As above	High – 2 f
Depression. Brown clay	Flooding; heavy texture and some waterlogging	N, P, K sufficient; pH satisfactory	No fertilizers required	Crop range more limited – rice, maize, cotton, fruit trees, sugar cane, tobacco	High to moderate – 2 f, t

Land unit	Physical limitations	Chemical status	Fertilizer requirements	Crop suitability	Rating
Levee over former river channel. Brown loam over sand	Flooding; low water holding capacity in subsoil	N, P, K deficient in subsoil only; pH satisfactory	N, P, K fertilizers required	Crop range more limited – fruit trees excluded	Moderate – 3 f, w, n
Former river channel. Brown sand	Flooding; low water holding capacity	N, P, K deficient; pH satisfactory	N, P, K fertilizers required	Unsuited for cultivation – limitations too great	Low – 4 f, w, n
Former river channel over depression. Brown sand over clay	Flooding; low water holding capacity in topsoil	N, P, K deficient; pH satisfactory	N, P, K fertilizers required	Unsuited for cultivation – limitations too great	Low – 4 f, w, n
River terrace Terrace surface – dense woodland and thicket. Red and brown sand	Vegetation clearance difficult; low water holding capacity but high water table	N, P, K sufficient; pH satisfactory	N fertilizer needed if maize and cotton grown	Cashew, sesame, sorghum, millet, cowpeas, cassava. Possibly maize and cotton	Moderate – 3 v
Terrace surface – open woodland. Red and brown sand	Low water holding capacity	N deficient; P sometimes deficient; pH satisfactory	N and P fertilizers needed if maize and cotton grown	As above	Moderate – 3 w, n
Depression – wooded grassland. (1) Black clay	Seasonal flooding; heavy texture and waterlogging	N, P, K sufficient; pH alkaline in subsoil	No fertilizers required	Rice only	Moderate to low – 3 f, t, a
(2) Brown sand with induration	Seasonal flooding; shallow rooting zone	N, P, K deficient; pH satisfactory	N, P, K fertilizers required	Rice only	Low – 4 f, s, n
Major river valley – riverine forest. Brown sand	Severe flooding; vegetation clearance difficult	N, P, K deficient; pH satisfactory	N, P, K fertilizers required	Cultivation too risky due to flooding	Low – 4 f, v, n
Major river valley – grassland. Black clay	Severe flooding; heavy texture and waterlogging	N, P, K satisfactory; pH alkaline in subsoil	No fertilizers required	Cultivation too risky due to flooding	Low – 4 f, t, a

Limitation suffixes
Physical limitations
f   Flooding
s   Shallow rooting zone
t   Heavy texture and waterlogging
v   Dense vegetation hindering clearance
w   Light texture and low water holding capacity
*   Only those land units visited in the field have been listed.

Chemical limitations
a   Alkalinity – pH of 8·0 or higher
n   One or more of N, P, K deficient

such that rice is the only safe choice. In addition the dense woodland and the thicket land units have clearing problems and the major valley land units cannot be used due to very severe flooding.

In considering the soil chemical properties it becomes more difficult to generalize about the land use potential as the ranges of values for the main nutrients tend to be wide both for the flood plain and river terrace land units (Table 9.5). For example, on the flood plain the 15 cm N level for the levee unit, the 30 cm N level and 15 cm K level on the depression unit all range from low to satisfactory levels. However, one can state that the pH levels are entirely within the range of tolerance of the major crops grown in this area (rice, maize and cotton), and that the levee unit has sufficient P levels throughout. For the former river channel, the levels fall below satisfactory for all three nutrients. It is therefore difficult to make specific, quantified recommendations of fertilizer requirements for a particular land unit and the summary in Table 9.6 should be regarded as applicable to most but not all areas within the given land unit. The table indicates that in general only the former channel deposits require fertilizer treatment.

For the river terrace units also, N, P and K levels are extremely variable with no unit having sufficient levels throughout the range although the black clay soil and sands under dense woodland at 15 cm do have sufficient levels of all nutrients (Table 9.5). The crop recommendations for the terrace surface land units therefore reflect the general nutrient deficiency of these sandy soils, concentrating on low nutrient demanders, whilst crops such as maize and cotton would definitely require N fertilizer at least. The alkalinity of the black clay depression soil means that only rice can be grown.

Overall, the picture of land use potential which emerges in this area is of a flood plain with minimal physical and chemical limitations, capable of supporting a wide range of cash and food crops, in contrast to the river terrace with poorly structured and nutrient-deficient soils which can support only the least demanding crops.

Considering the question of likely flood elimination and mechanized irrigation this obviously would lead to decline of nutrient levels in the flood plain, although again information is scant as to the expected rates of decline. However, the unit variability shown especially of soil texture would be a problem in the evaluation for an irrigation scheme since water supplied should be carefully adjusted to soil texture. Also in the case of the river terrace sands, continuous cropping may mean that fertilizers will become necessary even for crops with very low nutrient demands. Here again the chemical variability shown would hinder the calculation of exact dosage levels.

To summarize the relationship between the land units' physical and chemical soil properties and the resulting land use potential assessment, it can be said that the former does provide an adequate basis for general crop recommendations and an indication of those soils likely to require fertilization. As such the land unit soil data are suitable for a semi-detailed mapping scale of about 1:100 000. However, variability within land units and even soil types is such that predictions of fertilizer dosage levels and crop yields could not be made even if the necessary crop–soil relationships were known.

9.6 The integrated approach re-assessed

This case study has shown that the terrain units identified have in some instances considerable soil variability and as such are of reduced value for land use planning.

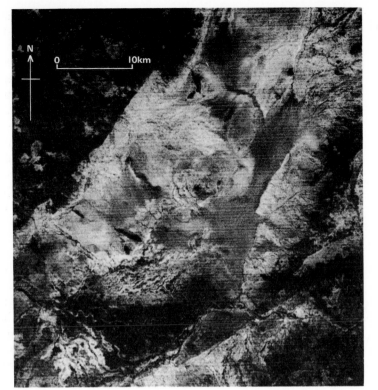

Figure 9.6 Landsat image of
the Mtera Basin, central
Tanzania (NASA Image).

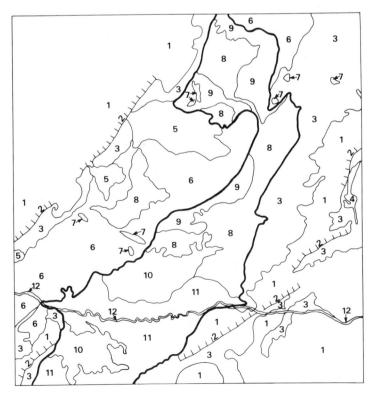

201

Is this a function of the approach used or of inherent environmental properties? Certainly for the flood plain, there is evidence to suggest that the unit variability is a function of the environment itself and independent of the survey approach. Anderson's (1961) conventional soil survey of the same area resulted in exactly similar mapping units, with one additional 'composite' unit of levee over depression not recognized in the present study. The chemical and physical soil data presented exhibited similar ranges of unit variability as found here. In Anderson's (1957) survey of the Mbarali irrigation area, a very intensive grid survey was carried out and a 1:10 000 scale map produced. Nevertheless cotton irrigation on the pilot scheme which followed has still been hampered by problems arising from soil heterogeneity (Watson & Tollervy 1971).

In the case of the river terrace land units, it is also doubtful that conventional soil survey would have eliminated the unit variability, in that this variability was shown to exist within soil profile types as well as within land units. For example, separate mapping of the red and brown sands would do little to improve unit homogeneity, since chemical variability was found to be considerable within these profile types.

It is therefore the conclusion of this study, that where limitations to the integrated approach have been revealed by field examination, these are a consequence of the environment itself. The integrated approach as such has been shown to be adequate for mapping the distribution of the main soil types, mapping units which do provide a basis for land use planning at the level of general crop recommendations and indications of likely nutrient deficiencies. It is important, however, to recognize that as yet this approach in this environment does not yield mapping units of sufficient homogeneity for more detailed planning needs, although problems of this nature have been shown to exist even with the most intensive ground soil surveys.

Looking towards the future, there is the possibility of improving the accuracy of integrated survey for soil mapping by employing other types of imagery and advanced interpretation techniques. To date, Landsat imagery has been used for land system/soil mapping of the Mtera Basin, central Tanzania (Cook 1977) where the flood plain land system could be readily delimited from the hills of the Basement Complex, the latter having a very dark-toned woodland fringed by very light-toned colluvial slopes as seen on band 5 (Fig. 9.6). Even at the small scale of 1:250 00 it proved possible to identify 12 land units as compared with the 17 derived from 1:32 00 air photos. More sophisticated interpretation techniques may have provided yet more information. Computer classification techniques have been tried on Landsat data as a basis for land system mapping in the Rukwa region of south-west Tanzania (Kirkula *et al.* 1978) although accuracy achieved was not good due to the spectral complexities of areas under subsistence cultivation. Nevertheless Landsat undoubtedly has potential as an additional data source for integrated soil surveys, and is certainly more efficient than air photos in time and possibly accuracy for the mapping of land systems. Its scope for land unit mapping has yet to be fully tested. This can also be said of other types of large-scale imagery such as multiband photography which could well yield more accurate land unit information than conventional aerial photography, but as yet such a sensing system has not been tested in an African flood plain environment.

References

Anderson, B. 1957. *Report on a soil survey of the Mbarali 5000 acre irrigation project.* Unpub. report, Dar es Salaam.

Anderson, B. 1961. *Report on the Rufiji Basin, Tanganyika.* Vol. VII: *Soils of the main irrigable areas.* Rep. no. 1269. Rome: FAO. *References*

Avery, B. W. 1962. Soil type and crop performance. *Soils and Fertilizers* **25**, 341–4.

Avery, B. W. 1973. Soil classification in the Soil Survey of England and Wales. *J. Soil Sci.* **24**, 324–38.

Baker, R. M. 1970. *The soils of Tanzania.* Unpub. report, Dar es Salaam.

Beckett, P. H. T. and R. Webster 1971. Soil variability: a review. *Soils and Fertilizers* **34**, 1–15.

Bie, S. W. and P. H. T. Beckett 1971. Quality control in soil survey. Introduction 1: The choice of mapping unit. *J. Soil Sci.* **22**, 32–49.

Bie, S. W. and P. H. T. Beckett 1973. Comparison of four independent soil surveys by air-photo interpretation, Paphos area (Cyprus). *Photogrammetria* **29**, 189–202.

Cook, A. 1974. *A photo-interpretation study of the soils and land use potential of the lower Rufiji Basin.* Res. paper no. 34.1, Bureau of Resource Assessment and Land Use Planning, University of Dar es Salaam.

Cook, A. 1976. *A photo-interpretation study of the soils and land use potential of the lower Wami Basin.* Res. paper no. 41, Bureau of Resource Assessment and Land Use Planning, University of Dar es Salaam.

Cook, A. 1977. *A soil survey of the Mtera reservoir area.* Res. paper, Bureau of Resource Assessment and Hand Use Planning, University of Dar es Salaam.

Gibbons, F. R. 1961. Some misconceptions about what soil surveys can do. *J. Soil Sci.* **12**, 96–100.

Gillman, C. 1949. A vegetation types map of Tanganyika territory. *Geogl Rev.* **39**, 7–37.

Kirkula, I. S., S. Kajula and R. B. King 1978. Assessment of computer processed Landsat data in areas of subsistence or no cultivation. Paper presented at *Conf. on National Surveys, automated Cartography and Remote Sensing, University of Durham, Dec. 1978.* Remote Sensing Society.

King, R. B. and A. Blair-Rains 1974. A comparison of ERTS imagery with conventional aerial photography for land resource surveys in less developed countries. Examples from the Rift Valley Lakes Basin, Ethiopia. *European Earth Resources Satellite Experiments,* Frascati, Italy. B. T. Battrick and N. T. Duc (eds), ESRO SP-100, 371–9.

Lawrence, C. J., R. Webster and P. H. T. Beckett 1977. The use of air photo interpretation for land evaluation in the Western Highlands of Scotland. *Catena* **4**, 341–57.

Mitchell, C. W. 1973. *Terrain evaluation.* London: Longman.

Mitchell, C. W. 1975. The application of Landsat 1 imagery to the Sudan Savanna Project. *J. Br. Interplanetary Soc.* **28**, 659–72.

Mulcahy, M. J. and A. W. Humphries 1967. Soil classification, soil surveys and land use. *Soils and Fertilizers* **30**, 1–8.

Parry, D. E. 1974. A natural resource evaluation of ERTS 1 imagery of the Central Afar region of Ethiopia. *Photogramm. Rec.* **8**, 65–80.

Parry, D. E. 1978. Some examples of the use of satellite imagery (Landsat) for natural resource mapping in western Sudan. In *Remote sensing applications in developing countries,* W. G. Collins and J. L. van Genderen (eds), 1–12. Remote Sensing Society.

Perrin, R. M. S. and C. W. Mitchell 1969. *An appraisal of physiographic units for predicting site conditions in arid areas.* MEXE rep. no. 1111, Christchurch, England.

Strömquist, L. 1976. Land systems of the Great Ruaha Drainage Basin upstream of the Mtera dam site. In *Ecological studies of the Mtera Basin,* 11–54. Stockholm: SWECO.

Trudgill, S. T. and D. J. Briggs 1977. Soil and land potential. *Prog. Phys. Geog.* **1**, 319–32.

Watson, J. S. and F. E. Tollervy 1971. Cotton research on the Mbarali irrigation scheme, Tanzania, 1957–1966. *Cotton Growing Rev.* **48**, 85–95.

Webster, R. and P. H. T. Beckett 1968. Quality and usefulness of soil maps. *Nature* **219**, 680–2.

Webster, R. and I. F. T. Wona 1969. A numerical procedure for testing soil boundaries interpreted from air photographs. *Photogrammetria* **24**, 59–72.

Wright, R. L. 1973. An examination of the value of site analysis in field studies in tropical Australia. *Z. Geomorph. NF* **17**, 156–84.

Young, A. 1973. Soil survey procedures and land development planning. *Geol J.* **139**, 53–64.

10 The role of remote sensing in mapping surficial deposits

*John R. G. Townshend
and Peter J. Hancock*

10.1 Introduction

Surficial materials are of considerable importance in several academic and applied disciplines. The term 'surficial (or superficial) materials' is usually taken to refer to the materials above bedrock whether they have arrived there by transport or have developed *in situ*. Investigation of the properties of surficial materials forms, therefore, a significant part of the studies of geomorphology and geology, particularly in the study of Quaternary deposits and landforms. Information about these materials also has considerable applied significance, as for example in the location of suitable materials for construction purposes, and location of suitable sites for structures such as buildings and communication lines.

A wide variety of methods has been developed to study surficial materials, from field-based estimates to highly sophisticated laboratory methods particularly developed by civil engineers and engineering geologists. However, especially in reconnaissance studies many of these tests are too time-consuming and expensive. For many investigations, remotely sensed data can provide valuable information to isolate the most suitable areas for a particular need, and hence reduce the amount of time spent in field and laboratory work. This is particularly true for regions which have poor geographic data bases. These include regions of rapid economic change such as the Gulf states, but the methods are no less relevant in lesser developed countries where financial resources are very limited.

In the first part of this chapter we examine how a physiographic approach to the study of remotely sensed images can readily lead to the subdivision of land into units useful for delimiting surface material types. In the second part we describe the application of digital techniques for classifying surface materials directly from remotely sensed data.

10.2 Physiographic approach to surficial materials discrimination

Belcher (1948) strongly advocated the use of air photographs as a source of information about the engineering properties of soils. Although his case was overstated (Hittle 1949, Vink 1968) air photographs are nevertheless still recognized as an essential source of information in themselves and as a guide to ground data collection. A notable recent example of their use was in the Bahrain surface materials resources survey (Brunsden *et al.* 1979, Doornkamp *et al.* 1980) which led

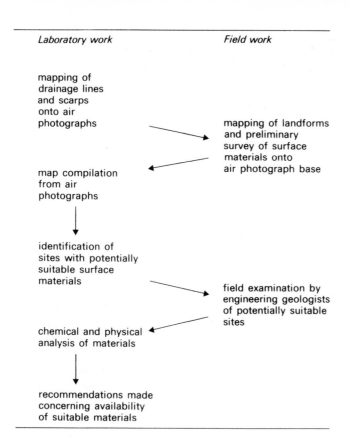

Laboratory work	Field work

mapping of
drainage lines
and scarps
onto air
photographs

mapping of landforms
and preliminary
survey of surface
materials onto
air photograph base

map compilation
from air
photographs

identification of
sites with potentially
suitable surface
materials

field examination by
engineering geologists
of potentially suitable
sites

chemical and physical
analysis of materials

recommendations made
concerning availability
of suitable materials

Figure 10.1 Main stages in a survey of superficial materials in the Bahrain Surface Materials Survey.

to the mapping of the whole of the island of Bahrain's superficial and solid geology at a scale of 1:10 000. The participants included geologists, geomorphologists (of which the first author was one), soil scientists and engineering geologists. The objectives of the survey were to locate materials suitable for construction and identification of suitable building sites, in particular those materials and sites free of the hazard of gypsum weathering, which in arid areas can rapidly lead to failures in foundations and lower parts of buildings constructed using concrete. This work illustrates a traditional approach to the use of remote sensing through air photographs. Figure 10.1 illustrates the main stages in the survey of surficial materials which formed a part of the overall survey. The approach concentrated on the use of geomorphological mapping. The first stage was basically morphological, simply comprising the mapping of drainage and major morphological features such as scarps. This formed a vital precursor to field mapping since it greatly facilitated ground location. The second stage was carried out in the field where a geomorphological map was produced with the basic character of the materials indicated (Fig. 10.2). Mapping the landforms was essential to make a first-order approximation of the areal extent of the materials. The importance of the air photographs varied considerably. In the case of the coastal plains called sabkhas which are produced predominantly by deposition, surface appearance is a far less reliable indicator of immediately subsurface materials than in the interior erosional areas and hence subsurface exploration largely through the digging of pits was necessary even at preliminary stages. A particularly significant role of air photographs in this survey

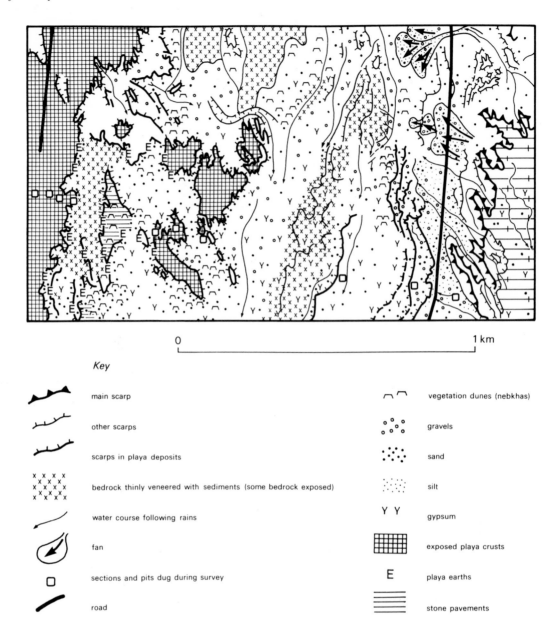

Figure 10.2
Geomorphological map
produced at a scale of 1:10 000
of part of the central
depression of Bahrain based
on black and white air
photographs and field survey.

was the provision of a detailed topographic base on which to map ground observations. Generally intensive field work was necessary because relatively little tonal variation was apparent on the black and white air photographs and because beyond short distances surficial materials maintained only a weak relationship with the landforms as displayed on the air photographs. Nevertheless an understanding of the operation of erosional and depositional systems combined with the spatial arrangement of landforms as revealed on the air photographs, were often important in later stages in assessing the value of surface materials. This is illustrated by considering the operation of an alluvial fan (Fig. 10.3). Clearly if the rocks eroded

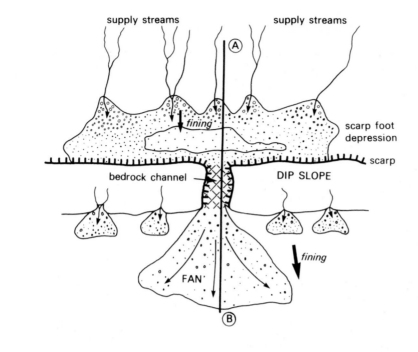

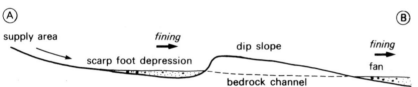

Figure 10.3 Schematic diagram of alluvial fan (Doornkamp *et al.* 1980).

by the headwater streams have an unsuitable chemical composition, in particular the presence of too much gypsum, the aggregate found in the alluvial fan is highly unlikely to have suitable chemical characteristics and hence detailed site investigations are unnecessary.

Work in several other regions strongly suggests that true colour stereo air photographs are more useful than conventional black and white photographs in unvegetated areas, whereas in vegetated areas false colour infrared photographs are more useful because they display vegetation conditions more effectively (Ch. 2) which in turn allows better forecasts of subsurface conditions (Section 4.7).

During the 1970s, space images have become increasingly important for mapping surface materials. This has included a survey of sand seas of the world using Landsat and Skylab imagery (MacKee & Breed 1974, 1976). Krinsley (1976) extended his previous work on playas in Iran (Krinsley 1970) using Landsat MSS data, to make recommendations for improved road location (Krinsley 1976). More recently acquired images from the Landsat 3 return beam vidicon cameras have a nominal resolution of only 24 m. Their value in permitting the recognition of landforms and surface materials is very clear in Figure 10.4a of the Senegal River, with its complex flood plain comprised of abandoned channels, oxbow lakes, and point bar scrolls. The limits of the flood plain are very distinct and broad areas of sand drifts can be seen to the north. Similarly in Figure 10.4b the contrasts between the thinly

207

Figure 10.4 Examples of the utility of Landsat 3 return beam vidicon images for the recognition of landforms and superficial materials. (a) Senegal River floodplain and partially sand dune covered landscape to the north. Area depicted 92·5 × 92·5 km. (b) Enlargement of part of the image (a), of the floodplain. (c) Playa and alluvial fans near Lovelock Nevada, United States. Area depicted 20 × 20 km. Image (a) contains caption information typical of Landsat images produced after 17 February 1977 (cf. description in Fig. 2.5).

02 JAN79 date of imaging; **C N16-26/W013-46** location of centre of image; **D** descending mode (cf. **A** for ascending); **219-049** path and row numbers indicating location of scene; **N N15-50/W014-24** nominal location of path-row centre; **R B** Landsat 3 RBV subscene (**A–D** corresponding to each individual MSS Landsat scene) or **MSS** sensor spectral band number (see fig. 2.5); **XA** code indicating RBV shutter duration, **O** code indicating presence or absence of aperture corrector; **SUN EL37 A137** elevation and azimuth of sun; **S** type of geometric correction; **2** size of area represented (viz. 92·5 × 92·5 km cf. 185 × 185 km for MSS); **S** type of projection (in this case space oblique mercator projection – others available); **P** data used to compute image centre (**P** predicted or **D** definitive); **N** normal vs abnormal (**A**) processing procedure; **L** low or high (**H**) gain; **NASA LANDSAT** agency and project; **E-30303-10400** scene identification number – as in Fig. 2.5, but with an extra digit for day numbers greater than 999 on photographic products created after 16 January 1978.

(a)

02JAN79 C N16-26/W013-46 D219-049 N N15-50/W014-24 R B XAOR SUN EL37 A137 S2S- P-N L2 NASA LANDSAT E-30303-10400-B

(b)

(c)

208

veneered rock surfaces and coalesced alluvial fans with coarse particulate materials and the playa surfaces are readily visible.

An indication of the degree of improvement in our knowledge of surficial deposits by means of space-altitude imagery can be illustrated using an example from western Argentina, to the east of the Andean cordillera, in the provinces of Mendoza and San Juan. The images used were true colour photographs obtained with Hasselblad cameras with a 150 mm lens from a high-altitude (230 km) sounding rocket (Drennan *et al.*1974). Resolution of this imagery was 25–30 m, comparable with the Landsat 3 R B V images or thematic mapper images (see Table 2.3). Additionally a small sample of true colour and false colour infrared air photographs taken within three weeks of the space imaging were available. Ground sampling was restricted to a traverse across the northern and southern parts of the area and thus the resultant map should not be considered a final statement, though it is unlikely that the broad outline of the features will be changed by subsequent work.

Figure 10.5a shows the surficial deposits as depicted on the 1:2 500 000 *Mapa Geologico de la Argentina* the best source available at the time of imaging. Using the space images, the map shown in Figure 10.5b was produced. It was not possible to use exactly the same categories on the second map as on the first. For example the distinction between loess and sand could not consistently be drawn. Nevertheless it is clear that our knowledge of the surface materials has been transformed by use of the space imagery. In particular, fluvial deposits are much more extensive than previously appreciated. On the very low angle slopes of this area it appears that very major changes in river direction have occurred. The surface materials are then modified by aeolian erosion and deposition. However, the river gravels beneath these remain *in situ*. Extensive fans also occur, notably between the rivers Tunayan and Diamante. The Tunayan itself has splayed out in an enormous inland delta. Large areas of ground are currently being deflated revealing highly reflective silt surfaces, which are akin to the much smaller 'wanderrie' named by Mabbutt (1963). Much more detail is potentially available from the photographs as is shown in Figure 10.6 of part of the Rio Tunayan, which was produced from true colour photographs through a 250 mm lens, resulting in a ground resolution of 13 m. Both contemporary flood plain and palaeo-flood plains have been identified, within which individual channels are readily located. Vegetated dunes are also found within the latter, but the main dune fields are to the south-west of the fluvial deposits. However, the latter are quite invisible on the true colour photographs obtained through the 150 mm lens, showing the significance of relatively small changes in resolution on the detectability of ground features.

All the examples quoted above have involved a physiographic approach, whereby the recognition of surface materials is based primarily on the recognition of landforms and to a limited extent use of land cover properties. Landform recognition is often dependent on the use of the stereoscopic model visible on air photographs, particularly when there is a substantial vegetation cover. Consequently the prospects for the mapping surface materials in humid vegetated areas using non-stereoscopic images from space altitudes will inevitably be limited, unless there is a particularly strong relationship between vegetation and underlying properties.

Key

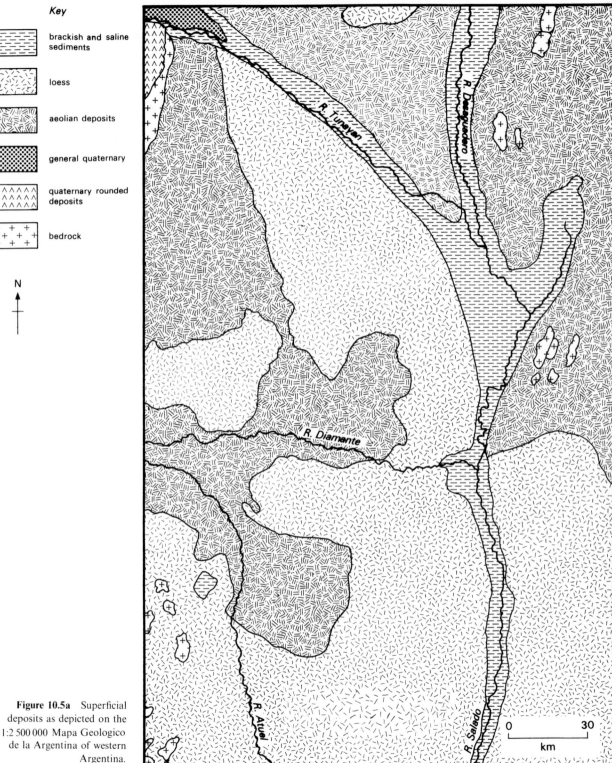

![brackish]	brackish and saline sediments
![loess]	loess
![aeolian]	aeolian deposits
![general]	general quaternary
![quaternary rounded]	quaternary rounded deposits
![bedrock]	bedrock

N

R. Tunayan

R. Desaguadero

R. Diamante

R. Atuel

R. Salado

0   30
km

Figure 10.5a Superficial deposits as depicted on the 1:2 500 000 Mapa Geologico de la Argentina of western Argentina.

210

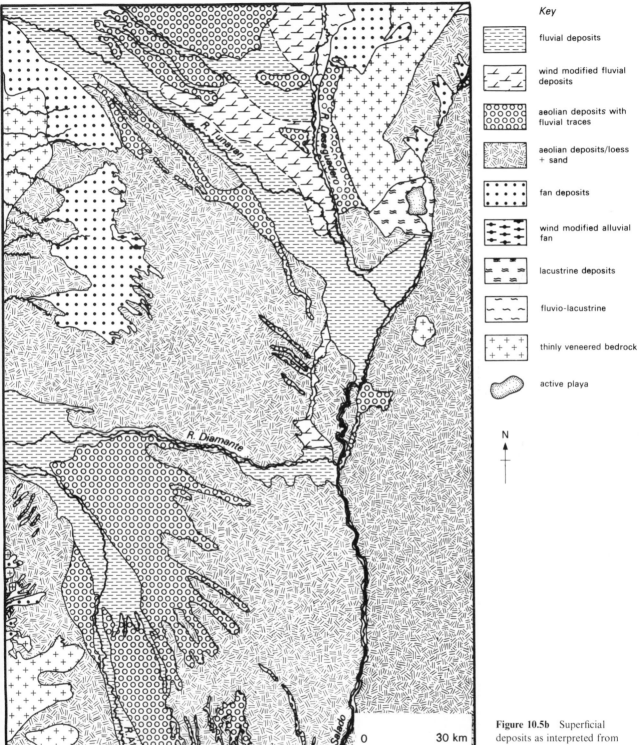

Key

	fluvial deposits
	wind modified fluvial deposits
	aeolian deposits with fluvial traces
	aeolian deposits/loess + sand
	fan deposits
	wind modified alluvial fan
	lacustrine deposits
	fluvio-lacustrine
	thinly veneered bedrock
	active playa

N

Figure 10.5b Superficial deposits as interpreted from Skylark Earth resources rocket photography.

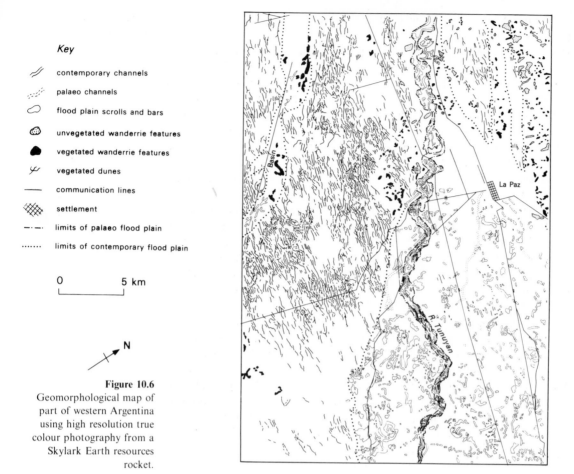

Figure 10.6
Geomorphological map of
part of western Argentina
using high resolution true
colour photography from a
Skylark Earth resources
rocket.

10.3 Quantitative digital approach

Despite the obvious merits of the physiographic approach relying on human visual interpretation, it is natural to question the extent to which some of the digital methods described in Chapter 4 can aid recognition of surficial materials. Clearly many of the image processing methods described in Chapter 4 can provide improved images more readily interpretable using a physiographic approach and some indeed were used in the above studies. More radically the use of complex ratioed images and their combination into colour composites by such workers as Goetz *et al.* (1975), have resulted in the enhancement and display of significantly increased information from Landsat multispectral scanner data.

The second author of this chapter has been investigating the extent to which bare coarse particulate surfaces can be discriminated using Landsat MSS data, using as a test area, the gravels of the flood plains in southern Italy of the Agri, Sauro, Sinni and Sarmento. These rivers contain within their catchments varying proportions of limestones, sandstones, clays, basalts and various metamorphic rocks, and these in turn contribute differing proportions of their respective materials to the bare particulate surfaces that comprise the flood plains. It was noted on photographic Landsat products of this region, that these bare surfaces displayed tonal variations, which, it was felt, should be attributable to variations in the materials present in the surfaces, though other factors such as vegetation cover and surface water area are also probably important.

A representative selection of these geological materials was sampled during ground data collection. To determine whether the different materials had distinctive spectral responses in the Landsat bands, the spectral response of each rock was measured using a band-pass radiometer (Fig. 3.4), with channels closely corresponding to those of the Landsat multispectral scanner (Milton 1979). Results from this laboratory investigation are shown in Figures 10.7, 10.8 and 10.9. The six material groups plotted are those that were most commonly found on the bare particulate surfaces. Other rocks such as granites and marls were also present, but in lower proportions. They would therefore, probably, have less effect on the overall signal received by Landsat MSS.

All three plots indicate that basaltic material, in terms of the three-dimensional spectral feature space, occupies a distinctive position at the very bottom of the spectral range. This is zone 1 that has been delimited on all the figures. Zone 2, on all figures, represents the area occupied by the iron concretions. At the other extreme to zones 1 and 2, lies the feature space that is occupied by quartzites and calcitic materials represented by zone 6 in Figure 10.7 and by zone 5 in Figures 10.8 and 10.9. In these zones there is some overlap of the pure white limestone and the quartzites and calcites. Between these two extremes lie the sandstones and the bulk of the limestones. Visually the former all appear redder than the latter. Therefore, one would expect channel 5 to be useful in their discrimination. Unfortunately, this channel alone cannot discriminate between the two types of material. However, in concert with channel 4 it is possible to delineate zones 3 and 4 (Fig. 10.9) that represent the distinctive feature spaces of the limestones and sandstones respectively. A minimal amount of misclassification has occurred during this 'zonation' process. It is interesting to note that the use of channel 5 and channel 7 reflectances (Fig. 10.8) failed to differentiate spectrally between the two material types. Thus zone 3, in this figure, represents their combined spectral feature space. Using channels 4 and 7 it is possible to divide the limestones into two groups; the first is more highly reflective than the second and occupies zone 5 in Figure 10.8; the second is darker and bluer in colour, with an increased percentage of chert in its make-up; this group therefore is relatively lowly reflective and occupies zone 3 in Figure 10.7. The sandstones occupy zone 4, that lies adjacent to zone 3, but because of higher reflectance in channel 7 lies above it.

It has been shown above that the selected geological materials can, and do, occupy distinctive positions in Landsat MSS feature space. It should therefore be possible to discriminate between surfaces that contain varying proportions of these materials. Work carried out by Krinov (1953), Hunt and Salisbury (1971) and Hunt *et al.* (1974) support the results given above.

Ground data were collected and Landsat MSS digital data were obtained that coincided with the ground data collection sites. Basaltic material occupies a very distinctive low position in the spectral feature space so this characteristic was used to aid discrimination of the surfaces.

The percentage of basaltic igneous material present in the ground sites was plotted against the reflectance in all four Landsat MSS channels. Channels 6 and 7 proved to have highly significant negative relationships (although R only reached -0.55). Channel 5 also displayed a negative relationship with this property, but unfortunately channel 4 gave little evidence of any relationship. Thus, in all channels, except channel 4, as the percentage of igneous material increased one could expect a decrease in reflectance.

213

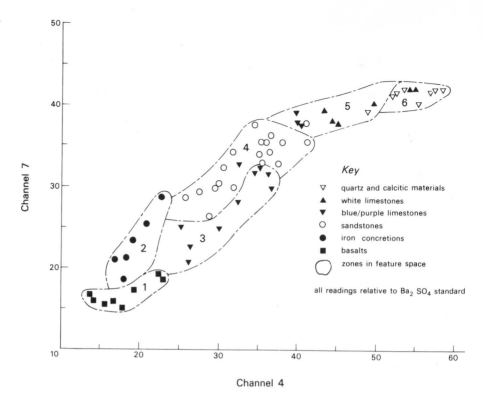

Figure 10.7 Plot of reflectance values of river gravels for channels 7 and 4 of the Landsat MSS obtained in the laboratory using a band-pass radiometer.

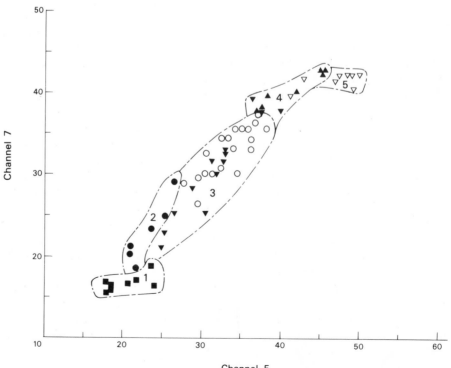

Figure 10.8 Plot of reflectance values of river gravels for channels 7 and 5 of the Landsat MSS obtained in the laboratory using a band-pass radiometer (see Fig. 10.7 for key).

214

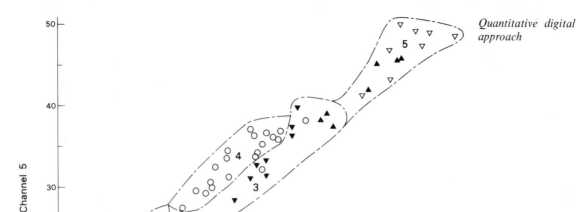

Figure 10.9 Plot of reflectance values of river gravels for channels 5 and 4 of the Landsat MSS obtained in the laboratory using a band-pass radiometer (see Fig. 10.7 for key).

The bare particulate surfaces were divided into groups with and without basaltic material. Since those surfaces without basaltic material are expected to be more highly reflective than those with basaltic material, the two groups should occupy different regions in spectral feature space. A discriminant analysis of the two groups was undertaken using the SPSS package (Nie *et al.* 1970). Using the spectral data available from the Landsat MSS, as the discriminating variables, the two groups were analysed. In this example 76% of the known cases were correctly classified. This is possibly as high as one could expect, considering the relatively low correlation values obtained between the surface properties and spectral reflectance.

The major assumption up to this point has been that different materials will produce different reflectance characteristics. To a certain extent, this has been proved by the laboratory work. However, there are undoubtedly other factors acting, such as particle size, that will affect the 'composite' reflectance spectra recorded by the Landsat MSS. Also, if the bare surfaces only display subtle spectral variations, these could easily be lost through errors in the sensing instruments or because of atmospheric effects. Moreover, natural surfaces are comprised of mixtures of materials which means that laboratory results will only rarely be matched by results from the field.

It follows from the above that the ability to distinguish between surface materials, using Landsat MSS digital data, depends on their spectral dissimilarity and the success of discrimination of surface materials depends on the mix of sediments present. Indeed, it is possible to obtain similar 'composite' spectra recorded by Landsat's MSS from particulate surfaces with different proportions of different materials.

215

Thus one requires a knowledge either of the particulate surfaces one is investigating, or of the geological materials contributing to that surface, as a precursor to detailed analysis. Landsat digital data alone could produce misleading results.

10.4 Conclusions and future work

Presently available satellite imagery has been shown to have value in surface materials discrimination, though for large-scale surveys more conventional data sources such as black and white air photographs will remain important. A recent investigation by the Laboratory for Applications of Remote Sensing (LARS) provides an interesting example of the integration of information from the two types of platforms. LARS has produced transparent maps of unsupervised classifications (Ch. 4) of Landsat data, registered to 1:25 000 maps to serve as an additional data source to conventional air photographs. Both sources are being used in the field by soil surveyors of the Indiana soil survey in an operational survey of the whole of Jasper County, Indiana (Weismiller *et al.* 1979), which is largely covered by glacial and fluvioglacial deposits and whose soils are largely a function of the geomorphological origins of these deposits. As before, the soil surveyors draw in the boundaries themselves, but it is hoped they will be located more readily and with greater accuracy than previously.

Future satellite systems will have much improved sensors for surface materials discrimination. Images from an airborne scanner with the same channels as Landsat D strongly suggest that much improved discrimination of rock types will be possible (Abrams & Rowan 1979, Podwysocki *et al.* 1979), and hence it is safe to assume this will improve the mapping of surface materials. On the other hand more recent work by Siegrist and Schnetzler (1980) indicates they are by no means optimal, even using current scanner technology. In particular a band centred on

1·25 µm was found to be best for discrimination and this is absent from the Thematic Mapper of Landsat D. Wider availability of thermal infrared imagery would also expand the role of remote sensing. Its value from aircraft has been demonstrated (Wolfe 1971, Sabins 1967, Watson 1971) (Fig. 10.10), but its cost inhibits more widespread use. Satellite imagery has had resolutions too coarse for most applications in the case of HCMM (Ch. 2) or of too limited availability in the case of Landsat 3. Thus we must wait for the thermal channel on Landsat D with its 120 m resolution before this part of the spectrum is properly explored and exploited.

References

Abrams, M. J. and L. C. Rowan 1979. Discrimination of altered rocks using spectral data from the 0·45 to 2·45 µm wavelength region. *Summaries 13th Int. Symp. on Remote Sensing of Environment, Ann Arbor, Michigan*, 7–8.

Belcher, D. J. 1948. The estimation of soil conditions from aerial photographs. *Photogramm. Engng* **14**, 482–8.

Brunsden, D., J. C. Doornkamp and D. K. C. Jones 1979. The Bahrain Surface Materials Resources Survey and its application to regional planning. *Geog. J.* **145**, 1–35.

Doornkamp, J. C., D. Brunsden and D. K. C. Jones 1980. *Geology, geomorphology and pedology of Bahrain*. Norwich: Geobooks.

Drennan, D. S. H., C. J. Bray, I. R. Galloway, J. R. Hardy, C. O. Justice, E. S. Owen-Jones, R. A. G. Savigear and J. R. G. Townshend 1974. The interpretation and use of false-colour infra-red and true colour photography of part of Argentina obtained by Skylark earth resources rockets. *Proc. 9th Int. Symp. On Remote Sensing of Environment, Ann Arbor, Michigan*, 1475–96.

Goetz, A. F. H., F. C. Billingsley, A. R. Gillespie, M. J. Abrams, R. C. Squires, E. M. Shoemaker, I. Lucchilta and P. Elston 1975. *Application of ERTS images and image processing and geologic mapping in northern Arizona*. Jet Propulsion Laboratory. Pasadena, Calif., Tech. Rep. 32–1597.

Hittle, J. E. 1949. Airphoto-interpretation of engineering sites and materials. *Photogramm. Engng* **15**, 589–603.

Hunt, G. R. and J. W. Salisbury 1971. Visible and near infra-red spectra of minerals and rocks. II: Carbonates. *Modern Geology* **2**, 22–30.

Hunt, G. R., J. W. Salisbury and C. J. Lenhoff 1974. Visible and near infra-red spectra minerals and rocks. IX: Basic and ultrabasic igneous rocks. *Modern Geology* **5**, 15–22.

Krinov, E. L. 1953. *Spectral reflectance properties of natural formations*. NRC Canad. Tech. Transl. TT–439 (translated by G. Belkov), Ottawa.

Krinsley, D. B. 1970. *A geomorphological and palaeoclimatological study of the playas of Iran*. Geol. surv. USDI, Washington, DC, AFCRL–70–0503.

Krinsley, D. B. 1976. Selection of alignment through the Great Kavir in Iran. In *ERTS–I a new window on our planet*, R. S. Williams and W. D. Carter (eds), US geol. surv. prof. paper 929, 296–302.

Mabbutt, J. A. 1963. Wanderrie Banks: microrelief patterns in semi-arid W. Australia. *Bull. Geol. Soc. Am.* **74**, 529–40.

McKee, E. D. and C. S. Breed 1974. An investigation of major sand seas in desert areas throughout the world. *NASA Goddard Space Flight Center, Proc. Symp. ERTS–1*, **1A**, 665–79.

McKee, E. D. and C. S. Breed 1976. Sand seas of the world. In *ERTS–I a new window on our planet*, R. S. Williams and W. D. Carter (eds), US geol. surv. prof. paper 929, 81–8.

Milton, E. 1979. *Reading bandpass radiometer user manual*. Remote Sensing rep. University of Reading, UK.

Nie, N. H., P. H. Bent and C. H. Hull 1970. *Statistical packages for the social sciences*, 2nd edn. New York: McGraw-Hill.

Figure 10.10 (*opposite*) Thermal infrared daytime image of the Thames valley around the village of Pangbourne, Berkshire, UK, located in the centre right of the image. The relatively cool water of the Thames enables it readily to be recognised. Darker bands can be readily seen on either side of the river showing the location of palaeo-channels within the flood plain. An extensive series of anastomosing palaeo-channels can also be seen in the valley leading towards Pangbourne from the area depicted in the top right of the image. Their darker tones relate to lower temperatures as a result of higher evapotranspiration rates arising from higher soil moisture, caused by differences in the composition of these surface materials. Knowledge of the location of such palaeo-landforms is usually of considerable assistance in the mapping of surface materials. Area depicted: approximately 6 × 3 km. Note geometric distortion caused by higher look-angle at margins of scan lines compared with the centre. (UK Crown Copyright, Ministry of Defence.)

217

Mapping surficial deposits   Podwysocki, M. H., F. J. Gunther, H. W. Blodget and A. T. Anderson 1979. A comparison of rock-discrimination capabilities based on present and future Landsat satellite sensor systems. *Summaries 13th Int. Symp. on Remote Sensing of Environment, Ann Arbor, Michigan*, 13.

Sabins, F. F. 1967. Infra-red imagery and geologic aspects. *Photogramm. Engng* **29**, 83–7.

Siegrist, A. W. and C. C. Schnetzler 1980. Optimum spectral bands for rock discrimination determined from aircraft scanner data. *Photogramm. Engng and Remote Sensing* **45**.

Vink, A. P. A. 1968. Aerial photographs and the soil sciences. *Proc. Conf. on aerial Surveys and Integrated Studies, Toulouse*, 81–141. Paris: Unesco.

Watson, K. 1971. Geophysical aspects of remote sensing. *Proc. Int. Workshop on Earth Resources Survey System*, NASA SP 283, **2**, 409–28.

Weismiller, R. A., S. K. Kast, M. F. Baumgardner and F. Kirschner 1979. Landsat MSS data as an aid to soil survey – an operational survey. *1979 Machine Processing of Remotely Sensed Data Symp., Purdue University, Indiana*, 240–1.

Wolfe, E. W. 1971. Thermal IR for geology. *Photogramm. Engng* **37**, 43–52.

1 Prospect: A comment on the future role of remote sensing in integrated terrain analysis

John R. G. Townshend

The remaining two decades of this century will bring yet further technological developments in the field of data collection by remote sensing and undoubtedly they will have major implications on terrain analysis. Many of the chapters in this book have looked forward to new sensors of the 1980s, such as the Thematic Mapper on Landsat D, and the sensors of the French SPOT system. Beyond the mid-1980s, the configurations of future systems are (at the time of writing) more problematic. Various speculations have been made in particular about satellite systems, some of which are summarized in Table 11.1, with indications of the feasibility of them being actually constructed in terms of foreseeable technological development. For many parts of the world, use of microwave radiation with its cloud-penetrating abilities seems necessary if the full monitoring capability of remote sensing is to be realized. But the high energy demands of such systems has slowed the installation of radars in satellites. Spacelab, hopefully, will be a valuable test bench for new sensors, enabling their development to be accelerated by provision of carefully controlled laboratory conditions in space.

Forecasting the character of remote sensing systems of the 1990s from the beginning of 1980s is a dangerous exercise but it is less difficult to propose some prerequisites for an increase in the contribution of remote sensing to the analysis of terrain. Firstly, it is important that basic research be conducted on the spectral, spatial and temporal properties of terrain characteristics themselves. Although substantial research has been carried out on the spectral properties of some crop types, there are many cover types that are far less well known, such as many tropical crops, forest canopies as well as urban surfaces. Moreover, the visible and near infrared has received the lion's share of attention and thus efforts need to be directed more towards other parts of the infrared and the microwave. More research needs to be performed on the bidirectional reflectance function, which describes the distribution of reflectance in relation to the direction of incoming and outgoing radiation. Such research is relevant firstly to the effects of terrain on reflected radiation, secondly to look-angle dependency of radiation properties as sensed from aircraft, and thirdly to the capabilities of new satellite sensors with variable look-angles such as the proposed Multispectral Resource Sampler (Schnetzler & Thompson 1979). In general our knowledge of the spatial characteristics of terrain attributes is much poorer than that of their spectral ones. There is need for basic work on the spatial properties of the phenomena themselves as well as the spatial characteristics of their reflected or emitted radiation, two sets often wrongly treated as synonymous. The wavelength dependency of the latter has as yet received little attention. It should go without saying that any sites for empirical investigations be extremely carefully chosen in order that the full range of types of spatial organization of terrain are included. The temporal properties of terrain

219

Table 11.1 Future satellite sensing systems (General Electric 1978).

Landsat H
(semi-credible)
	Smart optical sensors allow intelligent onboard editing/data reduction
	    forward/backward looking
	    10 m resolution, 10 bands
	    high resolution capability – 5 m targets (5 km)^2 scene.
	Synthetic aperture radar providing all-weather imaging capability: 25 m
	    resolution (3 band)
	Active visible sensor provides atmospheric calibration, luminescence and
	    night imaging selectable 3 km swath.
	Onboard processing and storage allows for change detection and/or
	    information extraction.
	3 spacecraft – 6 day repeat cycle

Earthwatch
(semi-credible)
	5000–10 000 km (6 h) repeating orbits provide near continuous Earth
	    coverage (8–12 satellites)
	Pointable optical sensors
	    high resolution for quick-look capability: 3 m resolution (5 km)^2, e.g. for
	    disaster assessment
	    medium resolution for mapping capability – 30 m resolution (90 km)^2
	    scene
	Microwave sensors frequency share antennae: 3 band synthetic aperture
	    radar – 10 m resolution; radiometer (passive microwave) – 3 band
	    12–120 km resolutions

GEOS (semi-credible to credible)	Large Earth-looking telescope for short-lived events with 24 m focal length mirror 8 m diameter primary optics: mirror segmented, adaptive controls Sensor images visible to thermal IR: 3 m resolution in visible 2D focal plane array of sensors
Texturometer (incredible to semi-credible)	Measures visible texture from 1 mm to 1 m from 600 km Mirror focal length = 600 m Adaptive optics for atmospheric correction, focus, pointing Data transformed to spatial frequency domain
Thermal inertia *mapper* (semi-credible to credible)	Follow on to HCMM (see Ch. 2) Measures thermal inertia or heat capacity of terrain Sequential passes over same area 4 am/10 pm – 4 pm/10 pm local crossings 10 m resolution in 8–13 μm band – 600 km orbit
Radar *holographer* (semi-credible to incredible)	Generates true VHF hologram Geosynchronous 300 mHz illuminator drifts 60°N + S of equator
Parasol *radiometer* (semi-credible to incredible)	High resolution soil moisture sensor with other applications Passive 10 km microwave radiometer From 1000 km orbit—10 m resolution at λ = 10 cm Phased array for sensing

Radar
ellipsometer
(semi-credible
to incredible)
	Bistatic radar approach uses one spacecraft for transmitter and one for
	    receiver (specular reflection only)
	Maps dielectric constants of soil and vegetation and vegetation height
	Measures effects of reflection on polarization; 600 km altitude, 1900 km
	    separation, 100 m resolution

Table 11.1 – *continued.*

Future prospects for remote sensing

Microsat	Primarily soil moisture sensor
	L-band passive radiometer
	Parabolic torus antenna
	Frequency 1·4 GHz
	Antenna approximately 600 m × 1300 m
	Ground resolution, 1 km; orbit, 1000 km
	Repeat cycle, 3 days (2 spacecraft)
	Radiometric temperature resolution, 1 K
Ferris wheel radar (semi-credible to incredible)	Large (30 km diameter) rotating (1 rev/h) cable structure relying on cable tension for support.
	Real aperture radar operates at low frequency (30–300 MHz) for ground penetration
	Resultant return signal can map materials (boundary layers and ground water) to depth dependent on soil moisture and salinity
	900 km orbit 300 m ground resolution
Sweep frequency radar (semi-credible)	Derives texture measure for identification and classification
	Polychromatic scatterometer from 30 MHz to 200 GHz
	Resonant backscatter indicates texture at discrete measurements from 1·5 mm to 10 m
	600 km orbit; 10 m resolution; 100 km swath
Geosynchronous SAR (semi-credible to credible)	Provides rapid update radar imaging capability
	System uses north–south drift of a geosynchronous spacecraft to provide range-rate for a synthetic radar
	Integrated signal produced by 8 minutes period of imaging
	Frequency, 2·5 GHz
	Ground resolution, 100 m
	Antenna size, 7·3 m

characteristics need to be established with attention paid to their variability from place to place and from time to time, in order to establish optimal dates and frequency of sampling. Such basic research on these three fundamental properties will help ensure that existing sensing systems are used sensibly in the investigation of terrain characteristics appropriate to their capabilities, and that in the future improved systems will be developed which are significantly better than current ones.

User requirements must be the ultimate arbiter in specifying the capabilities of operational systems. Technology in remote sensing has often preceded explicitly articulated needs for information. In the experimental and development stages this is quite appropriate, but not in subsequent operational ones. It is important that users of remote sensing data are encouraged, one might almost say coerced, to specify their data needs precisely in terms of temporal and spatial frequencies and spectral requirements. Even when such parameters have been stated, they are too often the result of intuitive speculations. For many tasks we already have systems which are needlessly overdesigned. Hence there is the retention of conventional black and white aerial photography for many purposes; other data collection systems may be potentially better, but not sufficiently better in terms of cost, available expertise or timeliness. Improved sensors may provide better resolution but have severe data handling costs; they may possess more spectral bands, but even these may not improve ground surface recognition; perversely such increases may worsen the results (Landgrebe 1978).

Associated with such developments in sensor and platform technology there is the need for improving the objective evaluation of the merits of images for specific terrain resources tasks: commonly called performance evaluation. In particular, the suitability of imagery in relation to the spatio-temporal characteristics of the terrain itself needs to be evaluated. Only in this way can appropriate remote sensing data and analytical methods be applied. Moreover, it should be possible *a priori* to make approximate predictions of the capabilities of sensing systems to provide data for given terrain analysis tasks, reducing the need for many of the seemingly endless evaluation exercises so common at present.

The usefulness of remote sensing data is of course also a function of methods of analysis and interpretation. Image enhancement techniques deserve attention, since human interpretation remains for many tasks the most efficient, cost-effective method in extracting information from remote sensing data. Better guidelines for choosing the most appropriate enhancement algorithm need to be developed as much as the development of new algorithms themselves. In the field of classification it is possible to identify several areas warranting attention. Newer methods such as context-classifiers and decision-tree classifiers need to be evaluated and further refined. Whatever classification algorithm is used the design of efficient sampling schemes for deriving training and testing samples is required. For many applications of terrain information it is becoming increasingly important not simply to classify the terrain but to derive quantitative estimates of terrain characteristics which can be fed directly into physically based models.

As pointed out in the conclusions to Chapter 4, sensible use of remote sensing data demands not inconsiderable acquisition of new skills of image processing and classification. Moreover, the most efficient use of newer digital techniques may require heavy investment in computing equipment and in output devices. This could provide a stumbling block both for those working in technologically less advanced countries and also for users with low budgets in other countries. In fact much can be done on relatively standard equipment and those providing educational courses in remote sensing must ensure that the technology is genuinely transferable and that they do not merely create unsatisfiable needs. We see therefore two important parallel lines of development in image processing and classification systems. Firstly, there is a need for the improved design of purpose-built image-handling environments for users to work within, both in terms of their physical and software components. Such systems must be readily accessible and exploitable by the genuine users of remote sensing data as well as the research workers. Secondly, at a more modest scale there must be education in the development and use of local facilities, using existing resources so far as possible. In education and training knowledge of the full range of possibilities should be disseminated. It is surely a mistake to teach only one facet of the topic, such as computer-assisted classification, unless potential users have had sufficient training to ensure their ability to assess its appropriateness for their problems.

Of the future developments in the handling of remote sensing data, none is likely to be more important than their integration with other data sources, to produce comprehensive geographic information systems (Section 5.3). The terrain analyst requires that all available information is brought together in spatially compatible forms: only then can a truly integrated approach be adopted. A further vital stage, outside of this book's scope, is that output from such geographical information systems must be produced in readily comprehensible forms for decision-makers. Unless significant effort is directed towards effective information display, little ultimately will be achieved.

In some respects, expectations of remote sensing are lower now than a few years ago. This is to be welcomed since earlier predictions were often overoptimistic and these have been replaced by more sober, carefully made judgements. The latter can only be made by the active participation of those directly concerned with terrain itself. The authors of this book all belong to this latter category and trust this book has provided a realistic assessment of the contribution of remote sensing to the study of terrain.

References

General Electric 1978. *Post Landsat D Advanced Concept Evaluation (PLACE)*. Space Division, General Electric, Valley Forge, Pennsylvania.

Landgrebe, D. A. 1978. Useful information from multi-spectral image data: another look. In *Remote sensing: the quantitative approach*, P. H. Swain and S. M. Davis (eds), 336–74. New York: McGraw-Hill.

Schnetzler, C. C. and L. L. Thompson 1979. Multispectral resource sampler: an experimental satellite sensor for the mid-1980s. *Proc. Soc Photo-instrumentation Engineers Symp.*, v. 183, Huntsville, Alabama, 255–62.

Appendix: some basic measures in remote sensing

Spectral resolution	describes the narrowness and position of the band of the electro-magnetic spectrum that is sensed.
Radiometric resolution	describes how precisely radiance is depicted in a set of data, as result of both the sensor and subsequent data transformations such as analog-to-digital conversion.
Spatial resolution	describes the fineness of detail that can be detected by a sensor.
Temporal resolution	describes how frequently a system senses an area.
Radiant flux	amount of radiant energy per unit time (units: watts).
Reflectance	the ratio of incident to reflected radiant fluxes.
Albedo	average reflectance over a given waveband, weighted by the distribution of incoming radiation in the solar spectrum. It thus represents the total radiant reflectance of natural objects.
Absorptance	the proportion of incident radiant flux which is transformed into another form of energy.
Irradiance	density of radiant flux incident on a surface (units: watts per square meter).
Exitance	density of radiant flux leaving a surface (units: watts per square meter).
Radiance	the intensity of radiant flux in a particular direction per unit solid angle per unit area of the extended source (units: watts per steradian per unit area).
Thermal inertia	measures the rate of heat transfer at the interface between two substances (units: watts per square meter per degree Kelvin second$^{1/2}$). Note that low thermal inertias mean there is high impedance to heat flow and vice versa.
Thermal diffusivity	indicates the rate at which temperature changes within a substance (units: square meters per second) and thus expresses the ability of a substance to transfer heat from or to its interior.
Dielectric constant	an electrical property of matter describing the ability of a material to store potential electrical energy.

Further explanation of these terms can be found in Reeves (1975), Monteith (1973), Nicodemus (1976) and Nicodemus *et al.* (1977).

References

Monteith, J. L. 1973. *Principles of environmental physics.* London: Edward Arnold.

Nicodemus, F. E. 1976. *Self-study manual on optical radiation measurements: part 1.* National Bureau of Standards Technical Note 910–1. Washington DC: US Department of Commerce.

Nicodemus, F. E., J. C. Richmond, J. J. Hsia, I. W. Ginsberg and T. W. Limperis 1977. *Geometrical considerations and nomenclature for reflectance.* National Bureau of Standards Monograph 160. Washington DC: US Department of Commerce.

Reeves, R. G. 1975. *Manual of remote sensing.* Falls Church, Virginia: American Society of Photogrammetry.

Author index

Abrams, M. J. 216
Abrosimov, I. K. 84
Adams, G. D. 51
Aggarwal, J. K. 61
Ahern, F. J. 145
Aitchison, G. D. 9, 124
Alford, M. 43
Algazi, V. D. 69
Allan, J. A. 135
American Society of Photogrammetry 59, 67, 68
Anderson, B. 187, 202
Anderson, J. R. 135, 138
Angelici, G. L. 98
Anuta, P. E. 141, 145, 146
Arsenault, H. H. 74
Aschmann, H. 133
Avery, B. W. 49, 184

Bagwell, C. 41
Baker, R. M. 185
Barnett, M. E. 62, 63, 74, 76, 77
Barrett, E. C. 18, 30
Batson, R. M. 67
Bauer, M. 39, 41, 44, 74
Beaumont, T. E. 68
Becht, J. E. 6
Beckett, P. H. T. 113, 184, 185
Beckman, N. 5
Beeman, L. E. 127
Beers, T. W. 44
Belcher, D. J. 38, 204
Benn, B. O. 5
Bennett, R. J. 11
Benson, A. S. 39
Bentley, R. G. 90
Berkebile, J. 146
Bernstein, R. 61
Bibby, J. S. 4
Bie, S. W. 184
Blong, R. J. 155
Bodechtel, J. 72, 134, 135
Bond, A. D. 89
Bonn, F. 39, 47
Bourne, R. 110
Bowman, I. 110
Brink, A. B. A. 110, 113
Brinkman, R. 6
Brooks, R. R. 84
Brooner, W. G. 89
Brueck, D. A. 119
Brunsden, D. 4, 204
Bryan, K. 155
Bryan, M. L. 47
Bryant, N. A. 91
Buol, S. W. 3
Buringh, P. 38
Burley, I. M. 8
Burns, K. L. 73
Buttery, R. F. 112

Caballe, G. 135
Campbell, J. B. 44
Canada Land Inventory 110
Carson, E. E. 4
Carter, L. D. 70

Cassinis, R. 134, 135
Centre National d'Études Spatiales 148
Chapman, P. 47
Chikishev, A. G. 110
Chou, M. 2
Christenson, J. W. 126
Christian, C. S. 5, 6, 51, 110, 111, 112, 115, 171, 175
Clarke, G. R. 49
Clawson, M. 8
Coates, D. R. 4
Cochran, W. G. 41
Coiner, J. C. 81
Cole, M. M. 51, 84, 90
Commission of European Communities (CEC) 135, 139, 140
Cook, A. 185, 186, 202
Cooke, R. U. 49, 111, 155
Cotton, W. P. 155
Cowan, D. J. 126
Cross, R. H. 127
Crossen, P. R. 3
Curtis, L. F. 40, 49

Dacey, M. F. 94
Dalrymple, J. R. 84
Dangermond, J. 127
Dansereau, P. 51
D'Auberade, E. M. 141
Davis, C. M. 127
Davis, K. P. 5
Demek, J. 49
De Sagredo, F. L. 135
de Schlippe, P. 1, 2
Dethier, B. E. 64
Donker, N. H. W. 62, 67, 69
Doornkamp, J. C. 204
Downs, S. W. 53
Draeger, W. C. 128
Drennan, D. S. H. 82, 90, 209
Driscoll, R. S. 50
Duda, R. O. 73, 86
Dudal, R. 10

Eardley, A. J. 84
Economy, R. 100
Electromagnetic Systems Limited 99
Ellis, S. L. 48
Environmental Systems Research Institute 127
Eppler, S. L. 53
Estes, J. E. 81, 82
Evans, R. 62
Everitt, J. H. 89

Farrow, J. B. 18
Fenneman, N. M. 110, 112
Fernandez, E. C. 135
Firey, W. 7
Fitzgerald, E. 31
Fleagle, R. G. 18
Flemming, M. D. 90, 92, 93, 145
Flouzat, G. 141
Fontanel, A. 136, 145
Food and Agricultural Organisation (FAO) 4, 6, 8, 9, 50
Fosberg, F. R. 50

226

228

Subject index

231

232

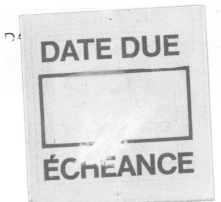